深圳市海阅通文化传播有限公司　主编

塑造商铺之王

Molding King of Shop:Commercial Design 商业店面设计 购物篇

江苏凤凰科学技术出版社

A site is just like a stage, which bears the development of the plot even though it can never be a role in any drama. However, actors get identities here while items gain significance, without stage, drama might be meaningless. So is shopping, it is not just a satisfaction on aesthetic needs and usage requirements, but more than stimulating personal desire to buy; it is more like a symbol of social identity and status.

Seemingly simple but actually complicate, consumption activity could be deepened a bit further. Even buying a same item, the significance of this item and the social status are much different in different sites. Each site has an accurate social identity with certain consumers, so the goods selling helps to deepen the self-reorganization and social identity. This is the key function of modern consumption market.

In that case, shopping sites are not pedlar's markets, but a certain place needs some kind of atmosphere to attract and appeal the consumers. In this site, consumers could find goods they need and feel the atmosphere to fit them, to handle their social status and search for social groups just like themselves. Entering this site, a consumer can explore or play his own role, and interpret a drama of certain stage of his life with others. This makes the site be more than a shopping site, but a site to create and exchange roles in the society instead.

Such a site is required to be high-end and gorgeous, and also aesthetic—being different from the flashy marketplace. A real atmosphere with a strong sense of aesthetics is called for, to be an accumulation of culture and key context for consumers to identify their identification. As a result, designs for those sites must be real designs instead of sticking to combination of luxury materials and inertia of thinking.

This book takes shopping sites as the theme, which fits modern consumption culture. What is more important, design for shopping sites must develop rapidly with the development of economy growth in our country. That is why we need to work hard and promote our design skills, to meet the need of overall industrial transformation in China.

PREFACE

场所，就像一个舞台。虽然它永远也不会成为戏剧的角色，但是总是承载着剧情的发展。演员在其中获得身份，道具在其中获得意义。没有了舞台，戏剧就会变得毫无意义。购物，不单纯是对使用需求和审美需求的满足，它还包含着个人购买欲望的宣泄，更暗示着买家的社会身份和地位。所购买的物品不仅仅是为了自己使用，更是在他人面前的一种展示。

这看似简单，实则复杂的消费活动可以有更进一步的深化。即使我们以同样的价格购买同样的商品，但只要消费场所不同，这些商品的意义和购买者的社会定位顿时会发生变化。不同的消费场所，吸引着不同的消费群体，做着准确的社会定位，同时以所出售的商品为标签更进一步加深了消费者的自我认同以及社会认同。这是消费社会中，消费场所的重要功能。

所以，购物场所不是集贸市场，它需要以某种气氛来感染、吸引它的消费者。在这个场所中，消费者可以找到自己所需要的商品，感受属于自己的环境氛围，寻找同层次的社会群体。进入这个场域，消费者或者寻找，或者扮演着自己的角色，与其他角色共同演绎着一场人生某个阶段的戏剧。当消费者从这个场域中走出去时，他们将带着这个身份奔向一个更加广阔的舞台。这类场所不仅仅是购物的场地，更是塑造角色并将其输入社会的一个身份转换地。

这样的场地不仅要高端、大气，同时更需要美感——不同于市井浮华的、真正的、带有强烈美学色彩的感受，这是一种文化的积淀，是消费者身份确认的重要内容。因此，这类设计不能再仅仅停留于华丽材料的搭配和惯性思维的驱使，必须走向真正的"设计"。

本书以购物场所为主题，恰好符合当下主流的消费文化。随着我国经济的发展，消费场所的设计必定迅速发展。设计界需要不断努力，提升自己的设计水平，以迎合国家整体的产业转型。

序言

CONTENTS 目录

Trendy Clothing 潮流服装

LIU·JO Store in Singapore 新加坡瑠久女装专卖店	008
Levis Icon Store Buttenheim 李维斯布滕海姆潮流时尚店	014
And A Kawasaki 川崎 And A 服装店	020
Appriva Shop Appriva 概念店	024
American Rag Cie Nagoya ARC 名古屋 服装店	028
Engelbert Strauss Work-wear Store 恩格尔贝特·施特劳斯工作服专卖店	034
Bulur Textile Moscow Shop 莫斯科布卢尔纺织服装店	040
Dikeni Men's Wear Boutique 迪柯尼男装专卖店	044
PUMA Brand Store, Osaka 大阪彪马旗舰店	052
L'Aurora Multi-brand Boutique L'Aurora 高级时装店	058
Inhabitant Store Tokyo Inhabitan 品牌东京旗舰店	064
MaxMara Hong Kong Central 麦丝玛拉香港旗舰店	070
La Koradior 珂莱蒂尔时装专卖店	074
Laurèl Flagship Store Hamburg 汉堡萝丽儿旗舰店	080
LIU·JO Collection Milan 米兰瑠久精品女装店	084
Degaje Showroom Degaje 服装展厅	090
Brooks Brothers Fashion Boutique 布克兄弟高级服装店	096
Zara Rome 罗马 Zara 概念店	102
Classic Precipitated Luxury – Giuseppe Custom Professional Service Center 经典沉淀的奢华——乔治白职业服高端定制	110
BUBIES Lingerie Flagship BUBIES 内衣旗舰店	116
Ecko Unltd 犀牛服饰	124
Hanna Trachten Flagship Store in Vienna Hanna Trachten 维也纳旗舰店	130
MASH Flagship Store in Beijing MASH 北京旗舰店	136
MUJI Shop in Milan 米兰无印良品	142
Ora Creation Store Ora Creation 概念店	148
Karl Lagerfeld Store, Paris 卡尔·拉格菲尔德巴黎概念店	156
Hemu Brand Store 荷木服装品牌专卖店	160
Ocean For in Huizhou 惠州四海名店	166

Shoe & Luggage 鞋、箱包

Cat Bag 猫猫包潮流店	172
Apple & Pie Children-shoe Boutique apple & pie 童鞋专卖店	176
Les Malles Moynat Les Malles Moynat 箱包店	182
Vintage Star – Jessica Simpson Collection 复古之星——杰西卡·辛普森品牌店	186
Munich Flagship Store Barcelona 巴塞罗那慕尼驰旗舰店	188

Jewel 珠宝

Wellendorff Jewellery Boutique 华洛芙高级珠宝店	196
Hirsh Jewelry Boutique, London 伦敦赫什珠宝精品店	200
Dinh Van Jewellery Boutique Dinh Van 珠宝精品店	208
Leo Pizzo Boutique Milan 米兰利奥皮佐精品店	212
Valuable Flagship Store, Suzhou 万宝缘银楼苏州旗舰店	216

Cosmetology & Hairdressing 美容美发

Pretty Beauty Health Spa 汕头唯美伊人养生美容馆	226
IL SALONE IL SALONE 美发店	232
Best Cuts and Colours Best Cuts and Colours 美发店	236
M'z Hair Design M'z Hair Design 美发店	242

Others 其他

Hao Jing Leisure 汕头豪景休闲会所	248
Qinglan Peninsula Club 清澜半岛会所	254
Casa VB Store Casa VB 生活专卖店	262
MUJI to GO Shop in Venice 威尼斯无印良品	268
Doctor Manzana Doctor Manzana 品牌空间	272
Whites Dispensary 白色药房	276
Masters Craft Palace Hotel Tokyo 巧匠工艺品店	282
Aleybo Flagship Store Aleybo 厨具用品旗舰店	286
Farm Direct Farm Direct 概念店	292
Retail Design Kabel Deutschland Kabel Deutschland 电缆零售店	296
SPAR Flagship Store SPAR 旗舰店	300
McIntosh AV Galleria 麦景图影音艺廊	308
O₂ Live Concept Store O₂ 生活概念店	314

CONTENTS 目录

款款服饰邀人赏
Various clothing longing for your admiration

绮丽华服，淡雅素裹，总是随着人的一举一动呈现不同的风情，灵动、随意、雅致、雍容还是云淡风轻，在这方寸间，自有安排。
Gorgeous costumes, elegant and simple dresses always give different feelings according to each move and act of the dressers. Even in such a small space, you can feel its brightness, causality, elegance and gentleness.

LIU·JO Store in Singapore
新加坡瑠久女装专卖店

Design agency : Studio Fabio Caselli Design
Location : Singapore

设计单位：Fabio Caselli 设计事务所
项目地点：新加坡

LIU·JO's new retail shop was located in Singapore's prestigious WISMA ATRIA Mall. The store's realization by Studio Fabio Caselli Design follows the Gold Concept's guiding principles of a harmonious combination of materials and space in order to represent its central concepts: Luxury and Made in Italy. Brightness, transparency, and precious materials are combined to create a space that is extremely feminine and enveloping.

A combination of gold and steel is employed as the decorations for the walls, columns and counters. The store's focal point is a semi-transparent screen, made of alternating cream and gold strips on steel structure, which divides and defines the surrounding space. Light is the fundamental element which binds and enhances all other elements employed in the Gold Concept, and on which great attention is placed in order to create a soft atmospheres that will draw the customers' attention and spark their curiosity.

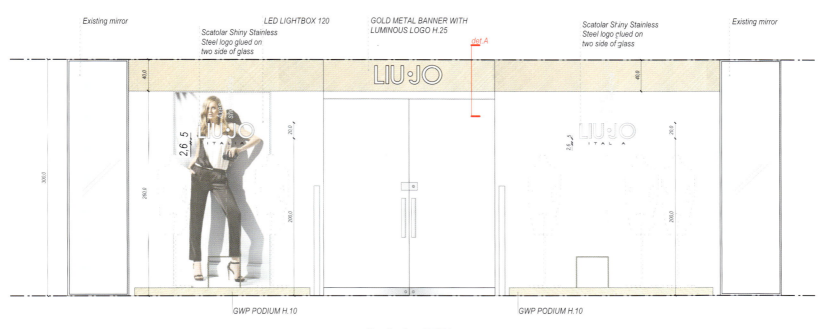

Elevation 1　立面图 1

瑠久品牌新的零售店位于新加坡著名的威士马广场购物中心。由 Fabio CaselII 设计事务所设计的商店实现了黄金概念的指导原则，材料与空间和谐融合，以表达其核心概念：豪华和意大利制造。亮度、透明度和珍贵的材料相结合，创造出一个非常女性化和充满围合感的空间。

钛金不锈钢用作店内展示架、展示台和整体空间的装饰品。店面中间由钛金色镜面不锈钢方管组成的隔离空间墙极具艺术设计感，也将相邻近的空间分隔开来。灯光的运用有效地强化了其他金属元素的设计概念，并为空间营造了柔美的氛围，吸引过路人的目光，激起他们进店一观虚实的好奇心。

Elevation 2 立面图 2

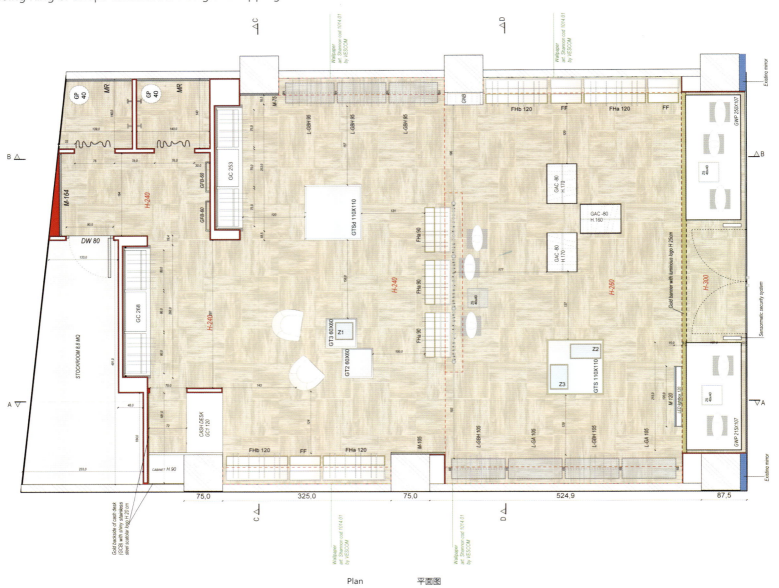

Plan 平面图

Levis Icon Store Buttenheim

李维斯布滕海姆潮流时尚店

Design agency : plajer & franz studio gbr
Location : Berlin, Germany
Area : 87 m²
Photography : die photodesigner.de

设计单位：plajer &franz 设计工作室
项目地点：德国柏林
项目面积：87 平方米
摄影：die photodesigner.de

The Levis icon store in Berlin is a small concept store designed by plajer & franz studio. The challenge was to redesign and rebuild the existing store in just 5 weeks before the Berlin fashion week.
The vision was to create a feeling of being in a typical old loft apartment with an industrial charm. Historical windows form shelf structures and window bars become cloth rails. Some glass panes are original and others are glazed with acrylic panes with Berlin "places to-be" printed on. Old radiators provide a basis for a presentation table which forms the center of the store and allows a varied presentation of products. The changing rooms and the cash desk were built with historic door elements. Levis merges with Berlin culture by using historic architectural elements in a quite artful way!

Cutaway View 1　剖面图 1

Cutaway View 2　剖面图 2

Cutaway View 3　剖面图 3

Plan　平面图

plajer & franz 设计工作室设计的新项目——李维斯潮流时尚店位于德国柏林。柏林时装周开幕在即,想要在五周的时间之内完成整个项目的设计和重建,这无疑是一个巨大的挑战。

本案的目的是营造出典型 LOFT 公寓的感觉,还需要带有些许工业气息。古老的橱窗被用作摆放商品的货架,橱窗上的木条则被用作挂衣服的杆子。部分玻璃窗保留原样,其他的有丙烯所作的画,如"柏林值得去的地方"等。一个古老的暖气片成为展示台的基底,放置在店内的中央,可以展示各式各样的商品。更衣室和收银台都是由旧门组建而成的。用这样一种巧妙的方式,将李维斯品牌与柏林文化完美地融合在一起。

And A Kawasaki

川崎 And A 服装店

Design agency: MOMENT
Designers: Hisaaki Hirawata, Tomohiro Watabe
Location: Kawasaki Kanagawa, Japan
Area: 167 m²
Photography: Nacasa & Partners

设计单位：MOMENT
设计师：平绵久晃、渡部智弘
项目地点：日本川崎神奈川
项目面积：167 平方米
摄影：Nacasa & Partners

One of the most important elements is the color scheme of red and black which are "And A" colors. The shop is divided into the three zones with the theme, 'future,' 'present' and 'past', where each zone is color coded relating to the themes. Each of them gives various sceneries in the shop. Facade made is attempted to impress passer-by and show a way to the future with a sharp and aggressive image. The slant line of the facade is the same angle as "A" of "And A," in other words; this also works as a billboard. In the center of the shop, there are dynamic elements such as distorted steel pipes and cake-dome display. The logo mark written with one stroke is given more motions. They lead to "present," with motions. Further, the shop draw the customers into "past" through an isolated wooden door.

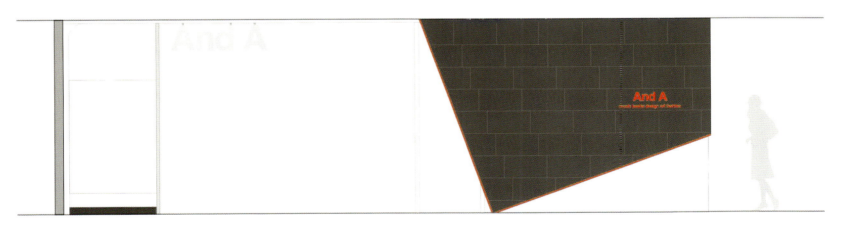

Elevation 立面图

店面外观与店内部空间设计装潢以红黑两色搭配为主。店内设计大体可分为三种主题区，分别是"过去"、"现在"和"未来"，在全新的设计下每个区都设计有特定的颜色和编码。黑色倾斜的三角形外观设计，形成店面的一大亮点；大面积的黑色面设计可更多地吸引顾客的视线，同时也象征着品牌名称中的"A"，如同一面广告牌在招徕过往的人们。也使店面角形外观轮廓更清晰、更出彩。店内的中央有一些动态元素，如扭曲的钢管和蛋糕层一样的展示台；品牌标志的运用使得店内更富动感。这些动态元素都象征着"现在"，另一个独立的木门则将顾客引领到"过去"。

Appriva Shop
Appriva 概念店

Design agency : Zemberek Design Office
Designer : Başak Emrence, Şafak Emrence
Location : Istanbul, Turkey
Area : 45 m²
Photography : Şafak Emrence

设计单位：Zemberek 设计事务所
设计师：巴塞克·艾伦斯、赛法克·艾伦斯
项目地点：土耳其伊斯坦布尔
项目面积：45 平方米
摄影：赛法克·艾伦斯

The project is a concept store, prepared for "Appriva" trademark of a textile firm, manufacturing for many trademarks. The area of the store is 45 m² and is located in the manufacturing facility of the firm in Istanbul. The design team aimed to create a location, which will support elements, such as high-end production and material quality and establish a connection between store and exhibited product.

In order to provide strong hints to the user concerning the indoors, the side of the store was designed using the materials of that time, such as wooden frames, shutters and elevated display shelves. Coating materials, such as wooden paneling, plaster panels with relief and mosaics are used, and integrity is provided in the perception of the era. All furniture in the space has been designed by the architectural team to comply with that era. The shelves in the upper part of the product exhibition units, are the deformation of shelves, frequently seen in retro train cars. The table in the middle is provided by pouring concrete in a special mold, made of brass. Brass mirror and coffee table were designed with inspiration from furniture, used commonly in 1950s.

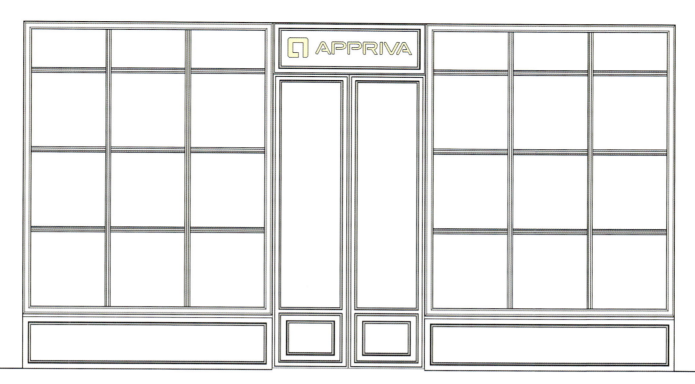

Elevation 立面图

本案是 Appriva 的概念店，它是一个专营纺织物的品牌。店铺面积仅 45 平方米，位于该品牌的伊斯坦布尔生产基地里面。设计团队想要建设一个能够维护高端成品和原料的项目，同时建立起店铺与展示之间的联系。

为了能让门外的人看到室内的景象，店铺的一侧使用了木质框架、百叶窗和高层展示架。覆盖材料如木质镶板、浮雕石膏板以及马赛克都被运用到项目中，打造出一种非刻意的时代感，毫无违和感。空间内的全部家具都由建筑团队精心挑选，与营造出来的年代感相契合。陈列架上的上部展示元件是变形的架子，这样的设计在复古的车厢里面能够见到。安放在店中央的黄铜材质的桌子，以及黄铜镜子、咖啡桌这些装饰都是 20 世纪 50 年代的流行元素。

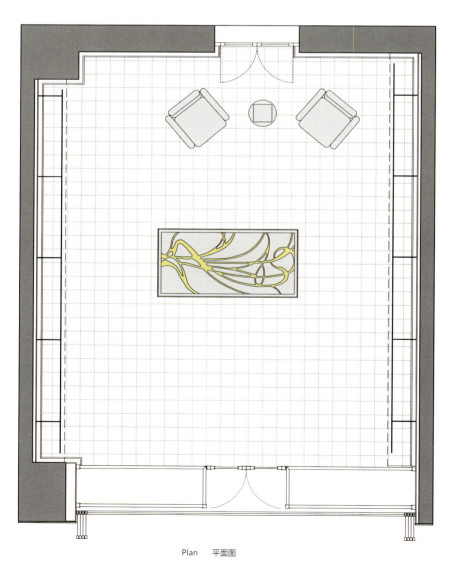

Plan 平面图

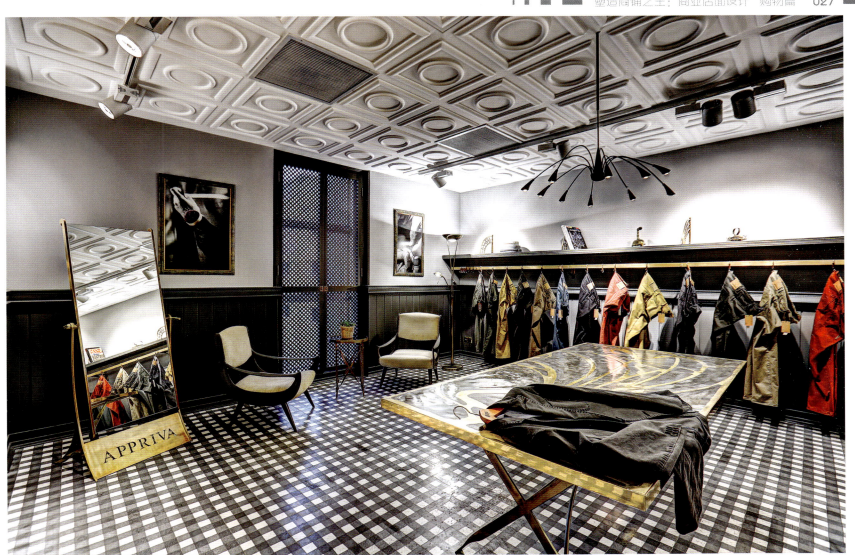

American Rag Cie Nagoya

ARC 名古屋服装店

Design agency : MOMENT
Designers : Hisaaki Hirawata, Tomohiro Watabe
Location : Nagoya, Japan
Area : 316 m^2
Photography : Nacasa & Partners

设计单位：MOMENT
设计师：平绵久晃、渡部智弘
项目地点：日本名古屋
项目面积：316 平方米
摄影：Nacasa & Partners

American Rag Cie is the multi-brand store where you can enjoy the women's wear on the first floor, and the men's on the second. It is attempted not to include the existing glazed building for this store design process, as the modern look always quickens passers by. It is important to slow their pace . The customers enter the store as if they go back to the past. The both floors are spacious as there is no partition, but only pillars. It softly divides the space; the arches on the ceiling signify the individual area. The intimate atmosphere is brought by the old time.

Elevation 立面图

美国瑞格是一个多品牌商店，一楼是女装，二楼则是男装。因为现代化的外观总是会让行人加快脚步，所以该项目并未保留原有的玻璃外立面，而是采用古风式的设计吸引路人并让他们放慢脚步。当客户进入店内，就仿佛置身于过去一样。两层楼空间都很充裕，没有隔断，只看得到柱子将空间柔和地分隔开来，天花板处的拱形图案象征着工业时代。如此一个毫无距离感的空间，将所有进入其间的人带回过去的世界。

1F Plan　一楼平面图

2F Plan　二楼平面图

Engelbert Strauss Work-wear Store

恩格尔贝特·施特劳斯工作服专卖店

Design agency : plajer & franz studio gbr
Location : Bergkirchen, Germany
Area : 2,400 m²
Photography : die photodesigner.de

设计单位：plajer & franz 设计工作室
项目地点：德国贝格基兴
项目面积：2 400 平方米
摄影：die photodesigner.de

The stores for Engelbert Strauss present the world of work-wear from a completely new perspective. The retail design concept is simple and unconventional but deeply honest. plajer & franz studio found a playful way to integrate various trade and craft references into the design concept and by doing so offer an emotional shopping experience with a clear expertise statement.

A real highlight for the youngest customers is certainly the bird's nest with an integrated slide! Deco elements such as carts transformed into display tables and mirrors, big car tyres used for the presentation of products as well as changing rooms built out of containers further enhance the emotional character of the working environment. Just as a side effect they achieve what is most important – a smile on people's faces! The shopping experience is complete only when you visit the Strauss café created from materials and elements of craftsmanship origin.

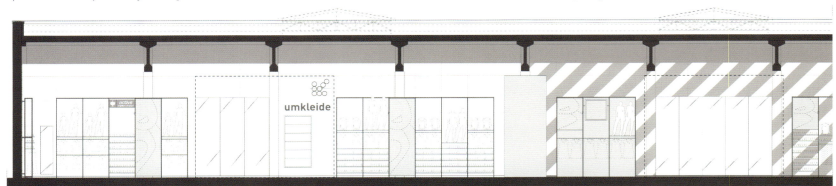

Elevation 立面图

Plan 平面图

恩格尔贝特·施特劳斯的服装店从全新的角度展示了一个工作服的世界。零售设计概念既简单又不落俗套，还能给人留下真实可靠的印象。plajer & franz 设计工作室找到了一种有趣的方法，将各种贸易和技艺参数融入该设计概念，以此用一种明了而专业的方式，为顾客提供一种情感购物体验。

店内最惹年轻人注目的亮点是一个集成片组成的鸟巢造型。像手推车这样的装饰元素都被拆分成为展示台和试衣镜，轮胎用来展示产品，集装箱还打造出神奇的试衣间，这些无疑都增强了室内环境的情感特性。与此同时，还收获了最重要的东西———人们脸上的笑容。只有从由表明工艺起源的原材料和元素构成的施特劳斯咖啡屋开始参观，才算得上是一次完整的购物体验。

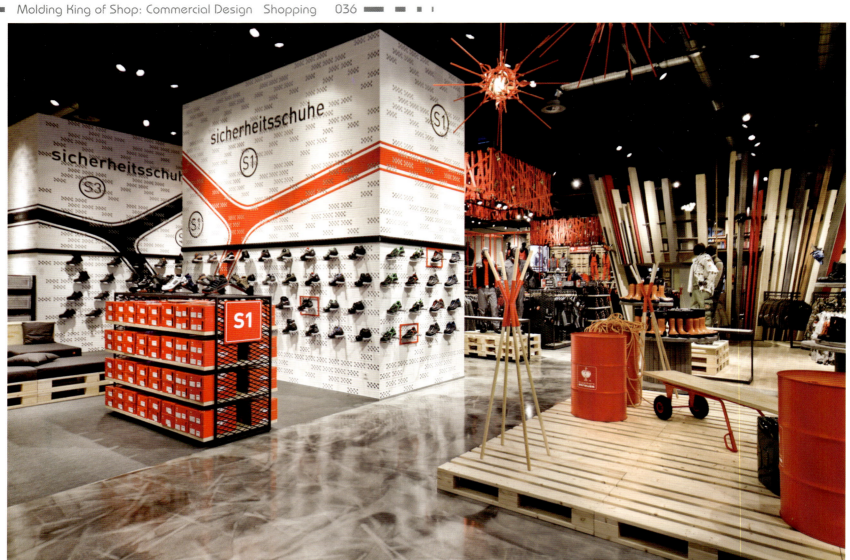

Bulur Textile Moscow Shop

莫斯科布卢尔纺织服装店

Design agency : Zemberek Design Office
Designers : Başak Emrence, Şafak Emrence
Location : Moscow, Russia
Area : 200 m²
Photography : Şafak Emrence

设计单位：Zemberek 设计事务所
设计师：巴塞克·艾伦斯、赛法克·艾伦斯
项目地点：俄罗斯莫斯科
项目面积：200 平方米
摄影：赛法克·艾伦斯

The project is a concept store, prepared for "Vigoss" trademark. The store with an area of 200 m² is located in a shopping mall in the city of Moscow in Russia. The design team has focused on a store design, the fact that denim jeans had entered nightlife from day use, and reflecting this issue on the quality of the product, as requested to be emphasized by the trademark. One of the targets, determined by the team in the process of formation of a concept, was to provide the feeling of a special design for the store and also, the ability to adopt the same to different stores, to give the same feel.

Continuous and uniform series of fronts in the corridor of shopping mall yield the necessity for the shops to emphasize their presences individually. With this thought, the front was withdrawn by sacrificing store area. Therefore, the store gave an inviting gesture in front of the user. Within the store, the total area was divided to different zones and the division of area to more well-defined zones, was provided. While this zoning allowed presentation of different collections, it also reinforces the perspective, presented by the space to the user. Denim wall, which has become a very dominant element of the space both within the space and with outside perception, consists of sheet metals, which are fixed with gussets, which are not in contact with each other. Wall units and middle modules, other than being mobile, have been designed with changeable heights for shelves and hooks, to allow exhibition of products with different sizes.

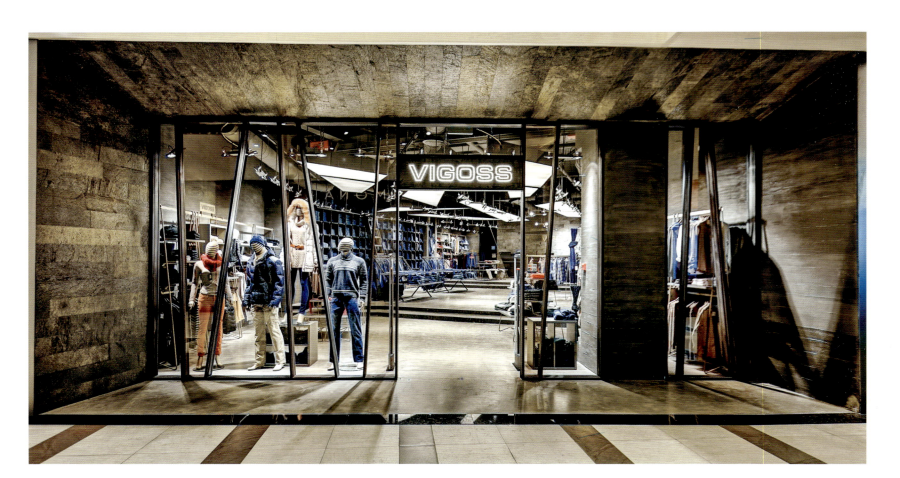

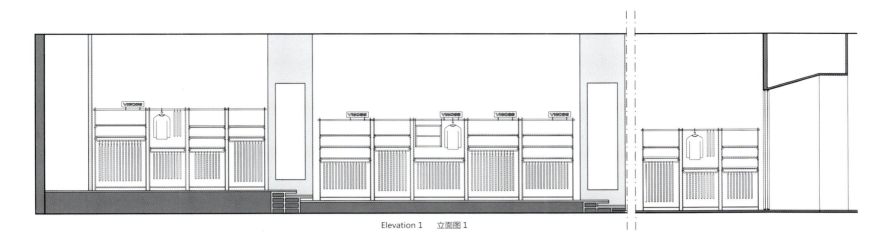

Elevation 1　立面图 1

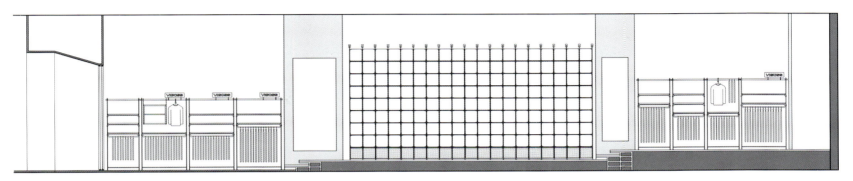

Elevation 2　立面图 2

本项目是为 Vigoss 牛仔品牌设立的概念店，面积约为 200 平方米，位于俄罗斯莫斯科的一个购物商场内。设计团队十分专注于店面设计，一个事实是牛仔已经从白天的衣着加入了夜生活必需品的行列，基于这一事实，产品的质量被要求给予强调。因此，设计团队的任务不仅是营造同类商店的相似氛围，更需要用特殊的技巧为其创造不一样的感觉。

在购物中心的走廊上有一系列连续的统一店门，所以对于店铺来说，要想在这一列门店中突出它们的存在是极其重要的。抱着这样的想法，设计师将前门往店内缩进，以一种邀请的姿态呈现在顾客的面前。踏入店内，你会发现空间被划分成了不同的区域，而在区域内也有着明确的分工。如此分区能够最大限度地展现不同产品的魅力，也为进入店内的顾客提供一个良好的视觉角度，进而挑选需要的产品。摆满牛仔裤的墙面是由被节点板固定并分隔成格子的金属板组成，它是影响内外空间感知的主导因素。墙面模块和中间的组件都是不可移动的，因此另外设计了可调高度的架子和挂钩，这样就可以展示不同尺寸的产品。

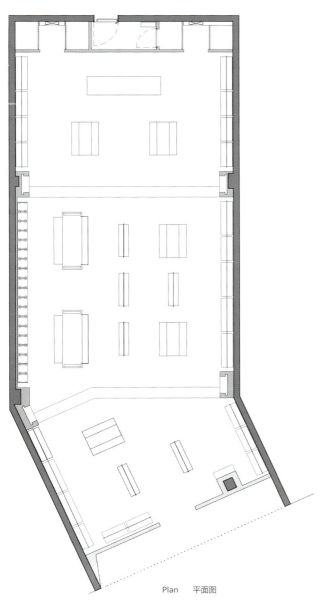

Plan 平面图

Dikeni Men's Wear Boutique
迪柯尼男装专卖店

Design Agency : Stefano Tordiglione Design Ltd	空间设计公司 : Stefano Tordiglione 设计事务所
Designer : Stefano Tordiglione	设计师 : Stefano Tordiglione
Location : Yingkou, Liaoning Province, China	项目地点 : 中国辽宁营口
Area : 1,000 m²	项目面积 : 1 000 平方米
Photos : Stefano Tordiglione Design Ltd	摄影 : Stefano Tordiglione 设计事务所

Spanning two floors, the project was a challenging undertaking, but its realisation is the embodiment of tastefulness and refinement. The storefront mixes shiny and matte metal strips in a dynamic manner to make the facade stand out, while the verticality of these lines makes it appear taller and more imposing. Large windows are interspersed with impressive marble columns and black drops that hint at the luxury and exclusivity of the store and its products within.

Elevation 1   立面图 1

Polished marble is a key feature of the first floor. Shiny black marble is used to create pathways through the shop's various sections, while off to either side grey and beige marble gleams, leading customers off to Dikeni's various products. Partitions break up the expansive space, made of glass or metal mesh to do so with a sensual subtlety. On the second floor, further products are displayed in a lighter setting. Large marble columns continue from downstairs but here the floor is of wood, lending a warmer ambience to this floor. There is a bar and seating area designed to be cozy and inviting thanks to lush sofas and warm wood.

这家旗舰店跨越两层楼，设计极具挑战。店面成为典雅和精致的化身，气派不凡。外立面由闪亮和亚光的金属饰条组成，极具动感，令人目不暇接，垂直的线条使店铺看起来更加高大、雄伟。特大的橱窗设计，点缀着别具特色的大理石柱和黑色吊饰，暗示了商店及其产品的高档与奢华。

抛光大理石是一楼设计的主要亮点。闪亮的黑色大理石通道将旗舰店的各个区域连接贯通，而通道两旁则闪烁着灰色和米色的大理石的光彩，引导顾客欣赏迪柯尼的各种珍品。由玻璃或金属网构成的区域将广阔的空间分隔开来，感觉细腻平衡。二楼相对于一楼比较轻柔明亮，展示各类产品。巨大的大理石柱从一楼开始，延伸至此。二楼也选用木地板，营造了一种更加温暖的氛围。楼上有一个咖啡区，豪华的沙发和温暖的木质结构为该区域营造了舒适宜人的气氛。

1F Plan 一楼平面图

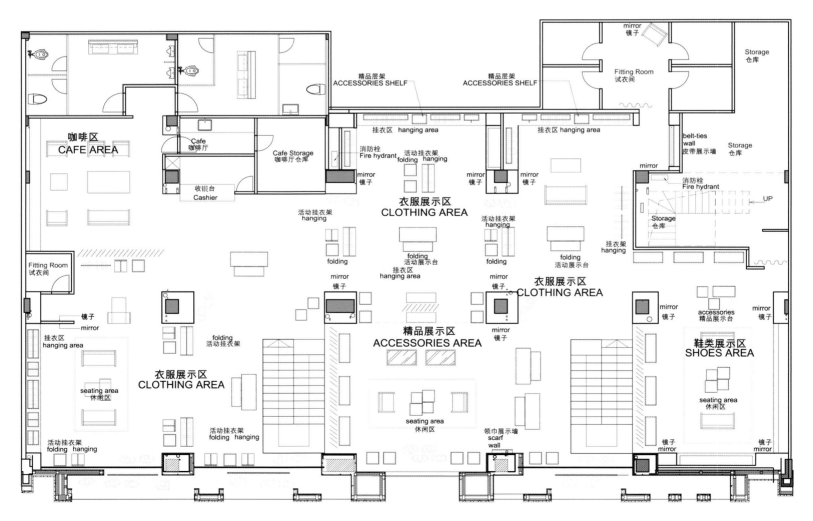

2F Plan 二楼平面图

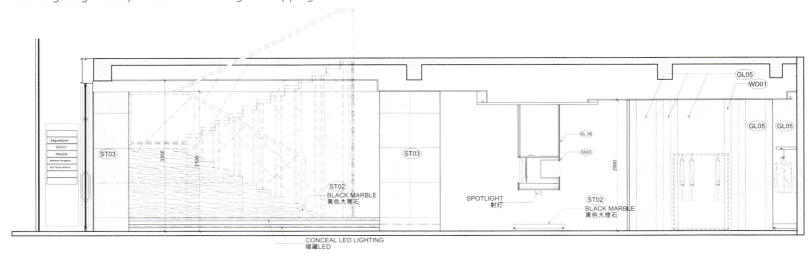

Elevation 2 立面图 2

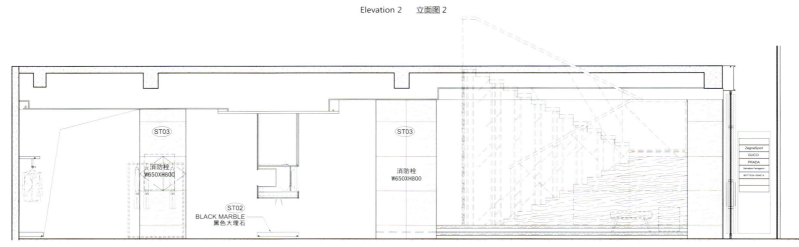

Elevation 3 立面图 3

PUMA Brand Store, Osaka
大阪彪马旗舰店

Design agency : plajer & franz studio gbr
Location : Osaka, Japan
Area : 579 m²
Photography : die photodesigner.de

设计单位：plajer & franz 设计工作室
项目地点：日本大阪
项目面积：579 平方米
摄影：die photodesigner.de

The PUMA is located in Osaka, Japan's flagship store, plajer & franz studio under the direction of Ales Kernjak (head of global store concepts, Puma Retail AG) developed a retail concept that is simple and flexible while integrating local references and offering PUMA's clients a joyful shopping experience. Black store in red alert for the ornament of color, this kind of classic black and red color collocation gives a person a kind of mysterious yet cheerful mood. Store product display shelves are designed like open style storage cabinet. Space colors overlapped ornaments, and the black as the theme of the store.
Thereby, the new premium the PUMA brand store located in Osaka is more than just a shopping place. It is a social and cultural meeting point, a space for events and happenings of various kinds. The lower two floors of the four-storey PUMA building have been designated for the 600 sqm shopping space, while the upper level – an open roof top, surrounded just by a light facade construction – creates an open space for performances and sport events. The façade is light, made out of meshed metal baffle allowing daylight to enter the store. At night, the interior lighting beams out, illuminating the street and giving a sneak preview of the shop's inside during closing hours. An impressive, cone-shaped staircase in the centre of the store with a red brand-wall in its back, is the eye-catcher making a strong brand statement.

位于日本大阪的PUMA（彪马）旗舰店，是由plajer & franz设计工作室设计的。设计师提出了全新的设计形式，建筑内外运用同质的设计语言及元素，并结合彪马公司的理念，打造一个简单灵活的零售设计，为彪马的客户提供一种快乐的购物体验，传播快乐和诙谐的运动精神。黑色的店内以红色为点缀，这种黑色与红色的经典色彩搭配，给人一种神秘却不失愉悦的感觉。店内的产品展示柜，设计上像是开放式的储物柜。空间色块与装饰物虚实叠交，和以黑色为主题的店面融为一体。

由此，新店不仅仅是一个购物场所，还是一个社会和文化的交汇点，一个购物娱乐和观看小型演出的空间。整个店有一个600平方米的购物空间，同时还创建了一个开放的空间作为演出和小型体育赛事之用。建筑的外墙被红色金属网所围绕，这样做的目的是为了能够让阳光照进店内。到了晚上，室内的照明又可通过网格照亮街道，马路上的行人便能看清店内的情形。建筑外墙上彪马的标志则被展示得十分醒目，设计师的目的是为了加强品牌形象的视觉传达。

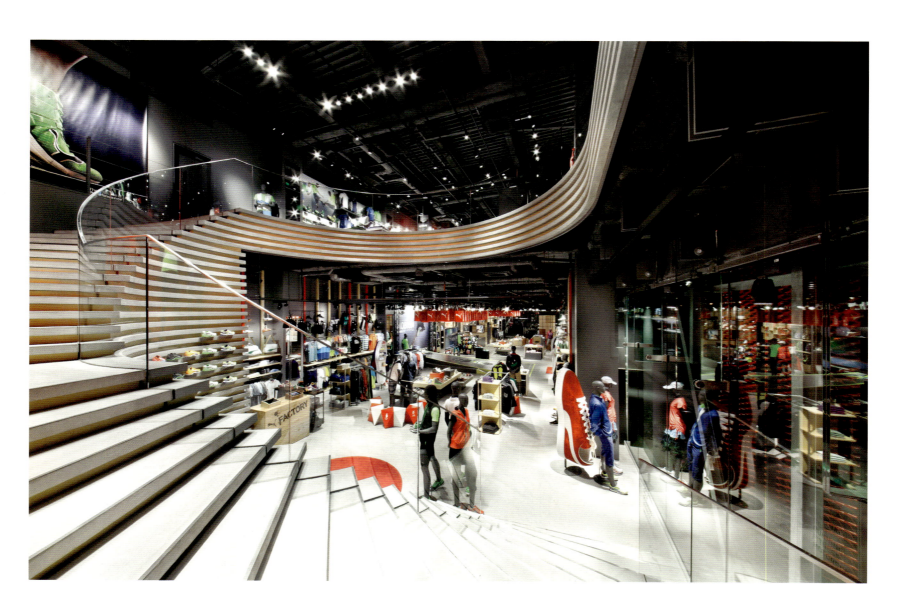

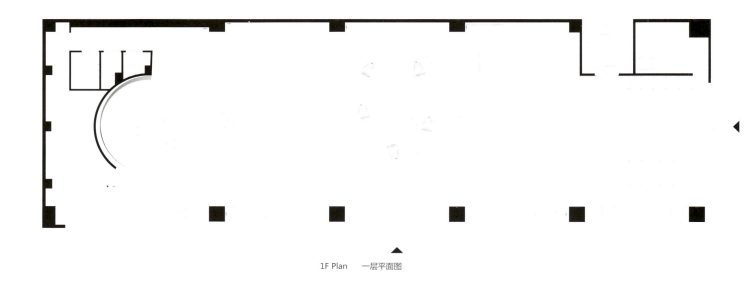

1F Plan 一层平面图

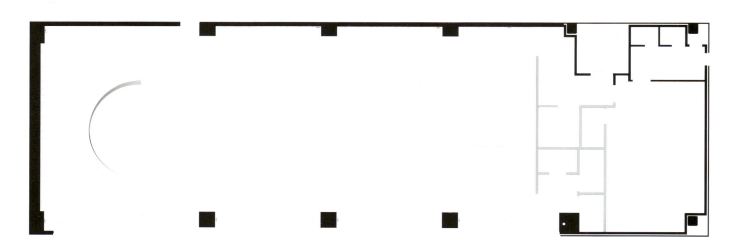

2F Plan 二层平面图

L'Aurora Multi-brand Boutique

L'Aurora 高级时装店

Design agency: Stefano Tordiglione Design Ltd
Location: Guangzhou, Guangdong, China
Area: 1,000 m²
Phototherapy: Stefano Tordiglione Design Ltd

设计单位：Stefano Tordiglione 设计事务所
项目地点：中国广东广州
项目面积：1 000 平方米
摄影：Stefano Tordiglione 设计事务所

From façade to fitting rooms, and everything in between, Stefano Tordiglione Design pulled out all the stops for the creation of L'Aurora at Happy Valley Shopping Mall in Guangzhou. The result was a striking yet harmonious design concept that spans the 1,000 square meter store, set over two floors.
Immediately the design of L'Aurora is impressive. Its façade references the works of Dutch painter Piet Mondrian, whose art features strong, dark lines and rectangular blocks of color. L'Aurora's storefront features pale hued and patterned glass, double-layered for greater impact, and set amongst bold lines. Inside customers experience more of a sense of the bold colors of Mondrian, through the designer's use – and design – of carpets with bright spots, swirls, prints and patterns and vibrant sofas that appear throughout the store to create a warm and welcoming, yet energetic vibe. This feeling continues throughout the first floor's women's wear section and upstairs to a second floor that features VIP rooms and men's wear collections.
The sprawling space is not only split up using intermittent seating areas but Stefano Tordiglione Design also makes use of passageways and intriguing hanging displays. Two large round birdcage-like areas are created via vertical metal poles. One ingeniously holds hanging rails for clothing that can be hung both inside and outside, while the other displays only shoes. Within more sofas create circles within circles.

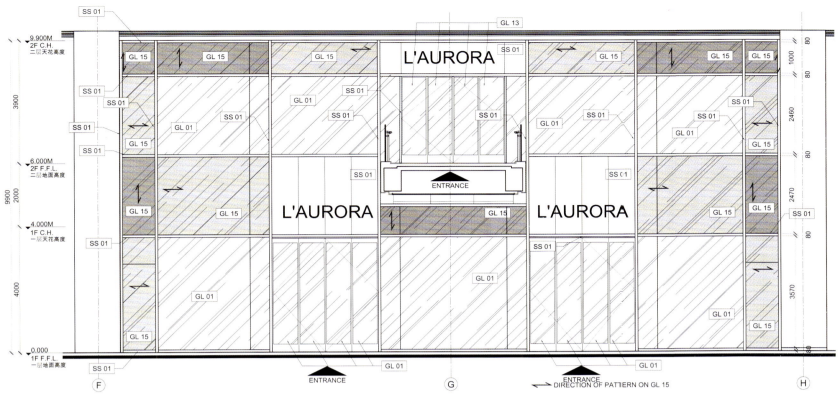

Elevation 1　立面图1

从外立面到试衣间，以及空间的一切，Stefano Tordiglione 设计事务所全力以赴，打造广州太阳新天地的 L'Aurora 高级时装店。这是一个引人注目的和谐的设计。时装店面积 1 000 平方米，跨越两层楼，令人感觉气派庄严。

L'Aurora 的设计，第一眼就能给人留下深刻难忘的印象。店铺外立面的灵感来自荷兰画家蒙德里安的画作，突出强劲有力的深色线条和彩色长方块。L'Aurora 外立面的特色是设置于粗线条中间的灰白色压花玻璃，其双层设计为顾客带来极大的视觉冲击。店铺内，通过设计师的巧妙设计，顾客可以体验到更多蒙德里安大胆的色彩搭配。布满亮点、涡纹、印花和图案的地毯和充满活力的沙发，遍布整个商店，创造了一种热情而又温暖舒适的氛围。这种氛围由整个一楼的女装区开始，并一直延伸到二楼的贵宾室和男装区域。

Stefano Tordiglione 设计事务所在对这个广阔的空间进行分隔时不仅使用了间歇式的休息区，还充分利用了通道和独特的悬挂展示。垂直的金属杆创造了两个大圆形区域，形似鸟笼。一个可以巧妙地焊接铰链横档，里外都可以悬挂衣服，而另一个只展示鞋子。空间里面，沙发画出了一个又一个圆圈，环环相扣。

塑造商铺之王：商业店面设计 购物篇

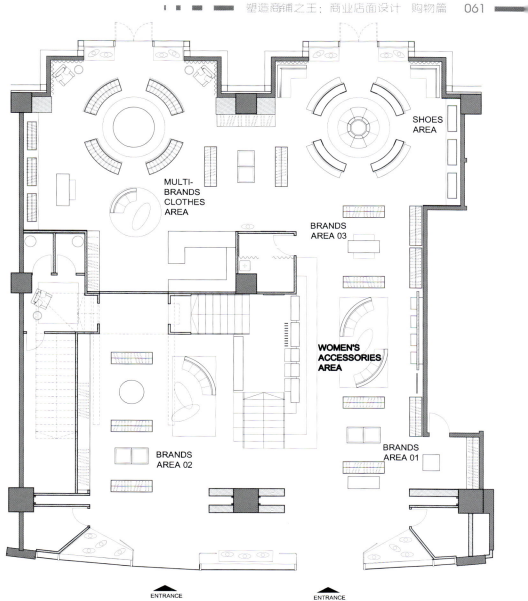

1F Plan 一层平面图

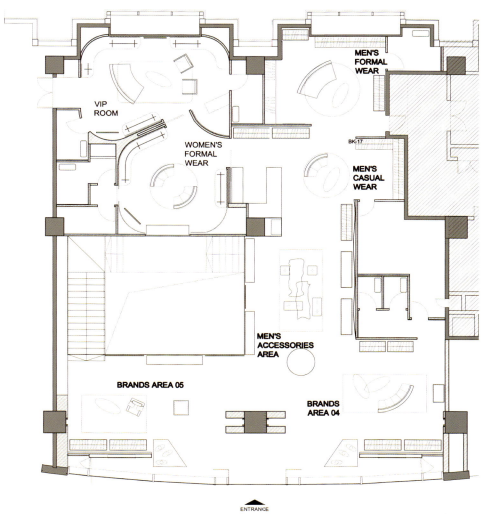

2F Plan 二层平面图

Inhabitant Store Tokyo

Inhabitant 品牌东京旗舰店

Design agency : Torafu Architects
Location : Shibuya Tokyo, Japan
Area : 225 m²
Photography : Daici Ano

设计单位：Torafu 建筑事务所
项目地点：日本原宿
项目面积：225 平方米
摄影：阿野太一

The Inhabitant Store Tokyo opened in Harajuku's Cat Street as the lifestyle/sport brand's flagship store. "Playfulness" and "Japaneseness" are the embodiment of Inhabitant's freestyle expression of modern Japanese taste which inspired us to envision a space thriving with the spontaneity of a casual stroll through the area known as "the back of Harajuku".

On each of the two floors, long plates cross diagonally the display areas with fitting rooms and counters positioned at a comfortable distance from them. Shoppers are greeted by a long plate on the first floor that can be used as a table to put articles on display, work or serve as a catwalk for special events. The edge of the plate becomes a step to the stairwell leading to the second floor where a suspended plate emerging from the wall welcomes customers like an overhead gate before extending diagonally into the display area holding hanger racks on its bottom side and multi-directional spotlights on its top to showcase the hexagonal tortoise-shell patterns spreading like clouds on the ceiling. Artist Asao Tokolo elaborated two patterns, whose every edge will always match each other, which can be seen encroaching on the floors, ceilings, walls and columns all over the store.

By capitalizing on the mutual relationship of the smaller units composing it, we strived to create a space that would in turn blend in with the small boutiques and residences that make up "the back of Harajuku".

1 CASHIER
2 SHOP
3 FITTING ROOM

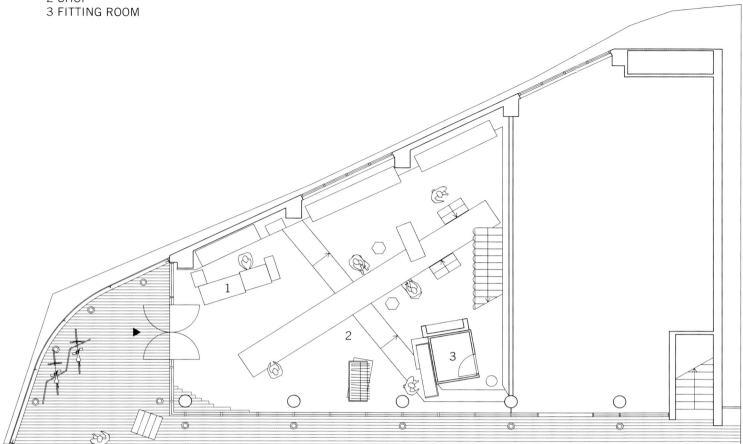

1F Plan 一层平面图

Inhabitant 运动品牌旗舰店位于东京原宿。其趣味性和日本特色体现出该品牌追求的自由精神，也正是日本年轻一代的生活方式。这启发 Torafu 事务所在远离各大时尚品牌店、遍布着各种个性商铺的原宿后街构建一个充满活力的空间。

在店面的两个楼层中，以对角线交叉分布的长板展示区域与柜台及试衣间保持着恰当的距离。在一层，长板作为桌子，可用于陈列商品、作为时装表演活动的 T 台。桌面的尽头同时作为通往二层楼梯的一节踏板。在二层，交叉线的形式变为天花板吊装，同样用于展示商品。六边形带有云纹图案的装饰出自艺术家 Asao Tokolo 之手，连续的云纹图案使这种以不同形式遍布于店内地面、立柱和天花的视觉元素可以无限延展。

Torafu 建筑事务所成功地创造了一个商业空间，它能自然地融入遍布各种个性商铺和住宅的原宿后街之中。

1 CASHIER
2 SHOP
3 FITTING ROOM
4 DISPLAY SPACE
5 STORAGE
6 PRESS ROOM

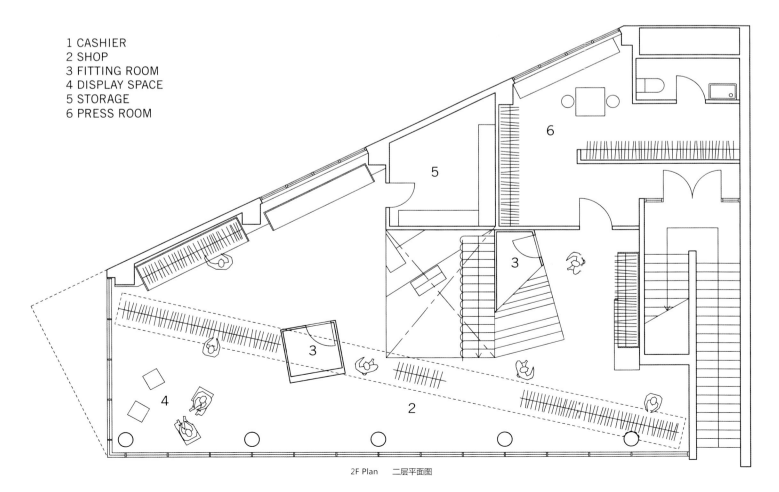

2F Plan 二层平面图

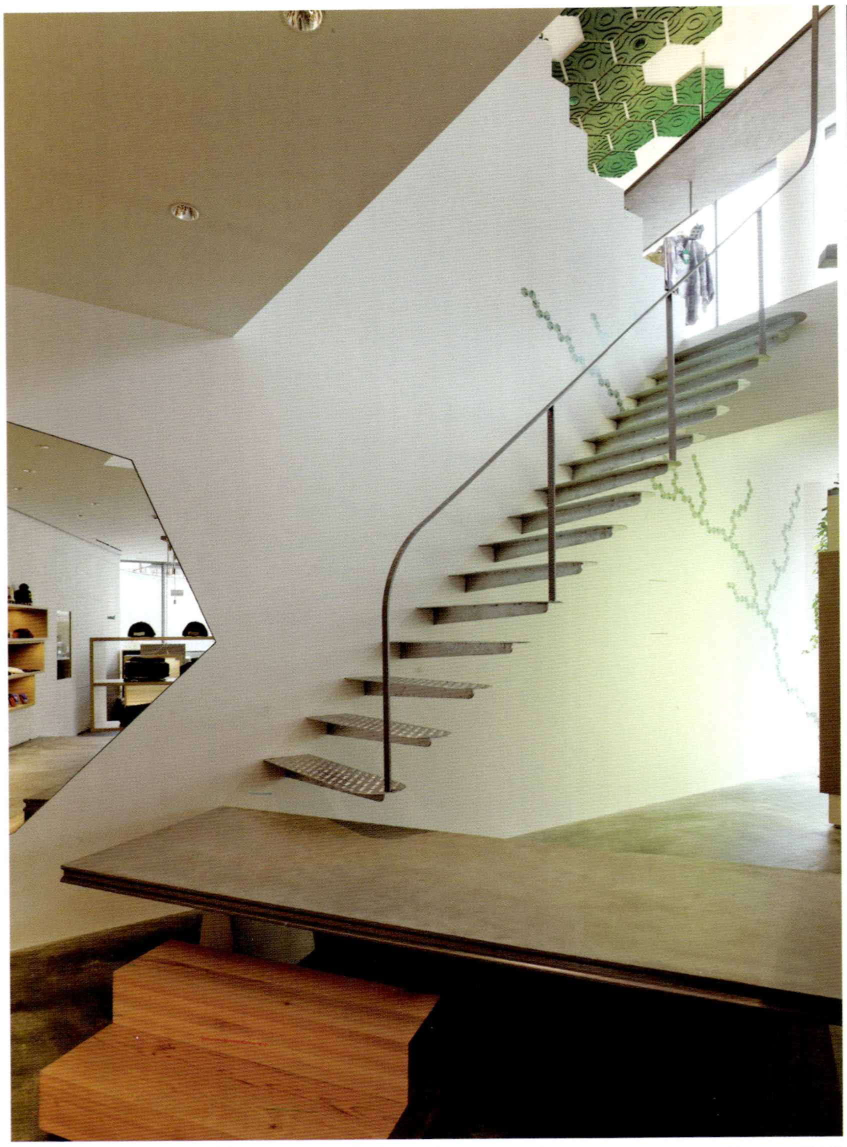

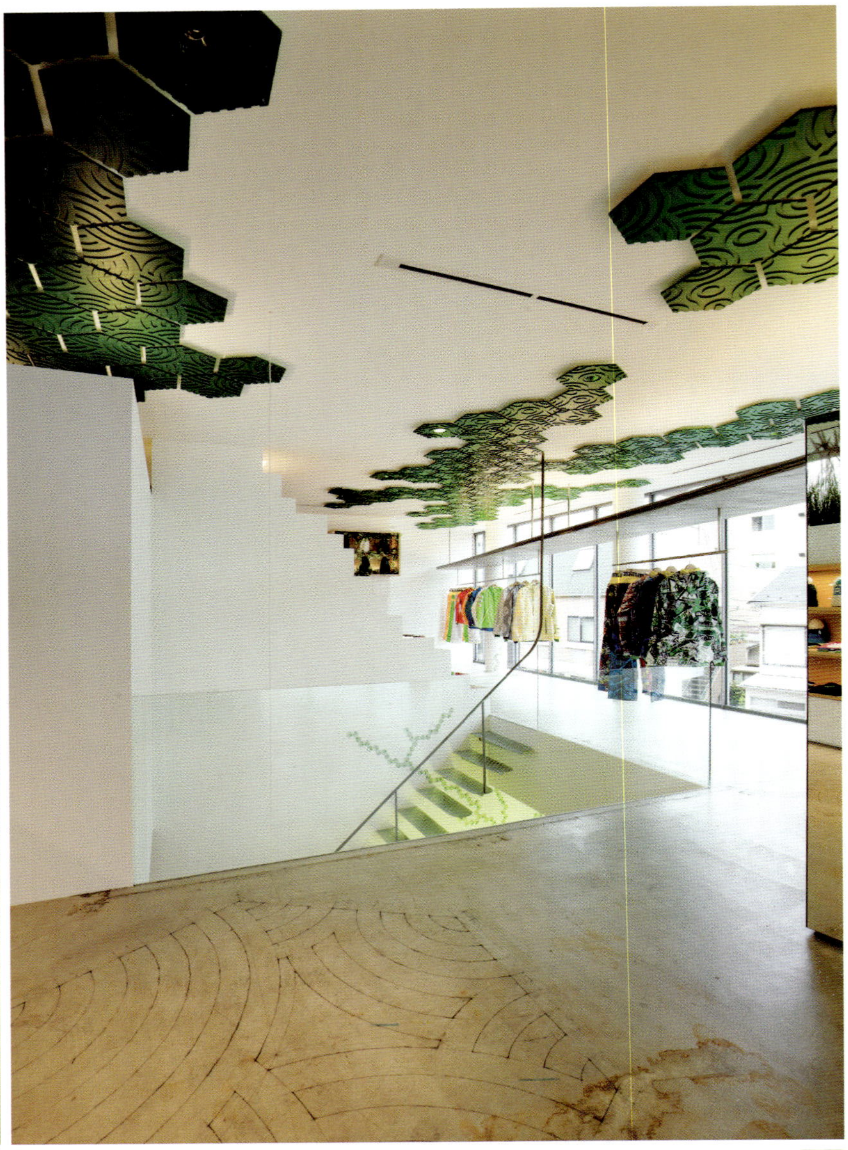

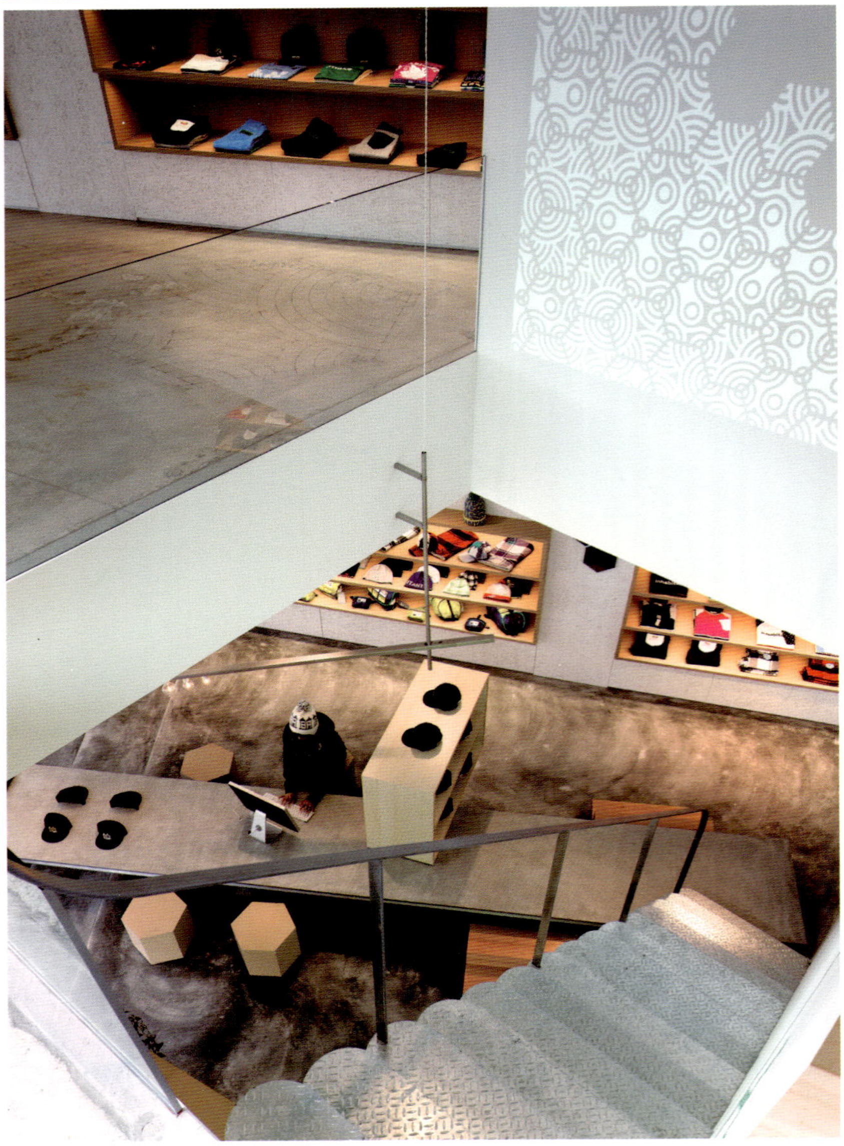

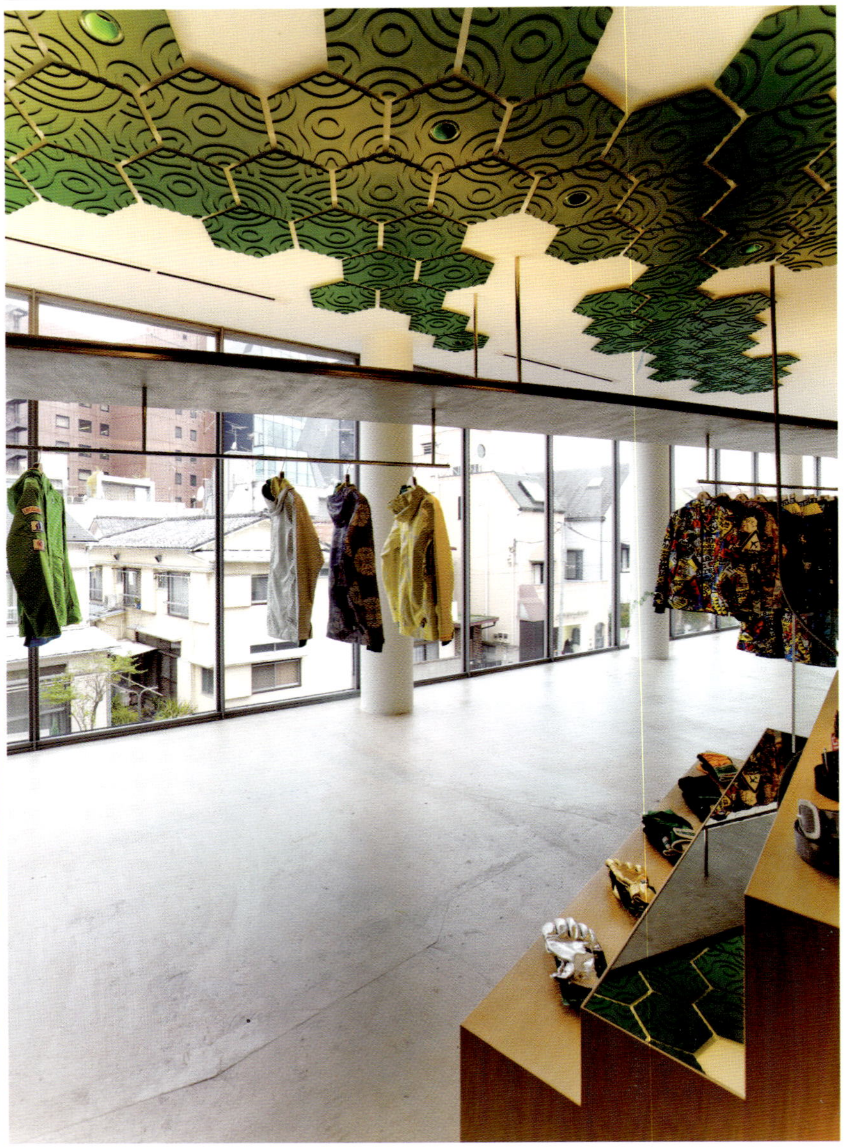

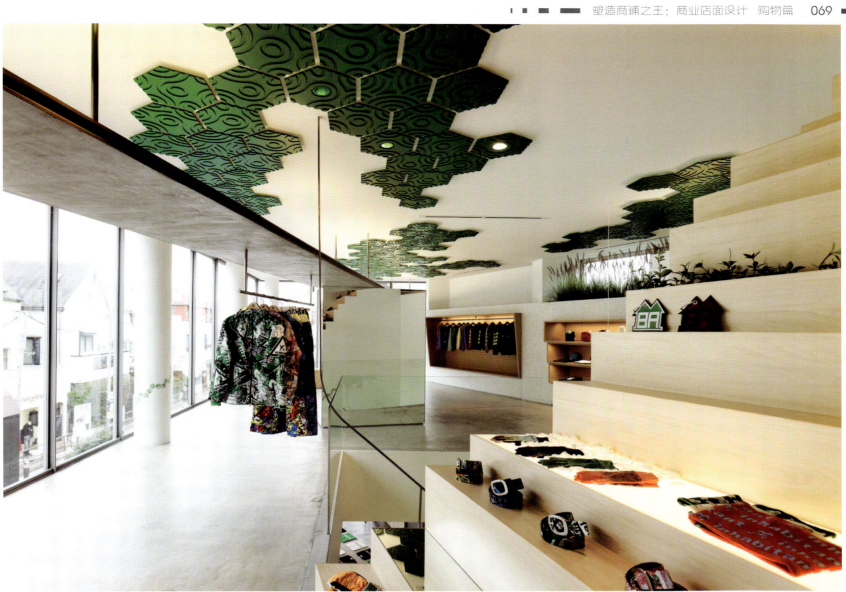

MaxMara Hong Kong Central
麦丝玛拉香港旗舰店

Design agency : Duccio Grassi Architects
Location : Hong Kong, China
Area : 450 m²
Photography : Virgile Simon Bertrand

设计单位：Duccio Grassi 建筑事务所
项目地点：中国香港
项目面积：450 平方米
摄影：维吉尔·西蒙·贝朗特

The new store in Hong Kong, located in one of Central district's main streets, financial and commercial center of the island, develops on two levels occupying a total of 450 square meters of surface in one of the most luxurious locations in town.

The two shop windows, which highlight the corner between Ice House Street and Chater Road, have a background consisting of two high definition LED screens, ensuring flexibility and reliability of communication. The wooden volumes that define the space and divide the fluxes recall Milan's store, while the big LED screens hidden behind a forest of curtains made of "midollino" sticks show natural images. To emphasize the hand-made and hand-craft character of italian production and MaxMara, it has been chosen to limit the use of cold materials such as stone or metal, preferring natural materials: ligneous essences characterize well-defined areas that form a harmonious whole both elegant and cozy. The walls are coated with these panels showing the impression of a kind of ancient fossil forest, creating a more private ambience where a living room, the elegant area and the shoes area are placed. The evolution of the concept admits unexpected combination between nature and technology, telling the charm of MaxMara's collection through screens integrated in the design of architectonic space.

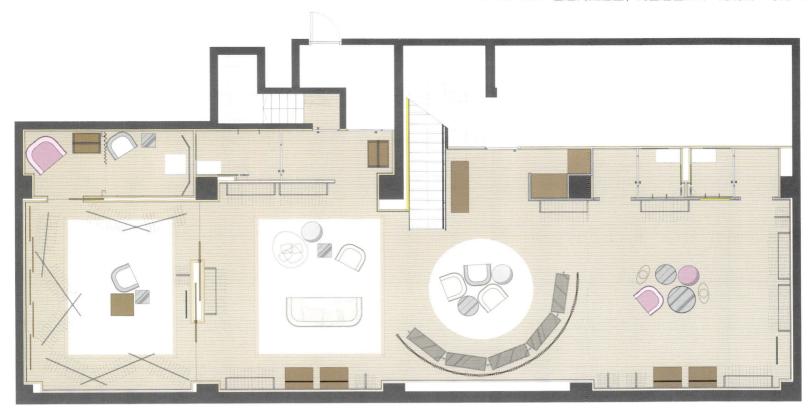

1F Plan　一层平面图

这一新店位于香港最重要的经贸区中心，在这样一个繁华的地段占据了 450 平方米的面积。

商店的两个橱窗突出了雪厂街和遮打道之间的角落，两个高清 LED 荧幕作为背景，确保了信息沟通的可靠性和灵活性。木质空间体量的使用和划分让人想起米兰旗舰店的样子，一个巨大的 LED 荧幕就隐藏在帘幕造就的深林里面，播放着自然景观的图像。为了强调意大利产品的工艺和手工特色，放弃了石材和金属一类的冷材料，店面更倾向于天然材料，使得分区极具特色，形成一个美观、舒适、和谐的整体。墙上装饰了一些木板，像一个古老的化石森林，内部设有客厅、鞋类展示区以及试衣间，创造了更为私密的空间。空间的演变体现了自然与科技的意想不到的结合，通过特别设计的屏幕，诉说着麦丝玛拉品牌的魅力。

2F Plan　二层平面图

La Koradior
珂莱蒂尔时装专卖店

Design agency : Stefano Tordiglione Design Ltd
Location : Wuxi, Jiangsu, China
Area : 100 m²
Phototherapy : Stefano Tordiglione Design Ltd

设计单位：Stefano Tordiglione 设计事务所
项目地点：中国江苏无锡
项目面积：100 平方米
摄影：Stefano Tordiglione 设计事务所

Entering La Koradior's boutique, the shop becomes the focal point. Customers are drawn to the walls having an elegant wave to them, immediately suggestive of the curtain that might have just drawn to close the first act. Lush materials cover large, soft chairs where customers can pause for a rest. The shined, stone floor is a mix of white, grey and coral-colored tablets. Deep red, rounded carpets highlight the central furniture while fitting in with a color theme of deep purple, used for the chairs, and highlights of cream and gold. An adjacent VIP room caters to the store's high flyers and here soft velvet furnishings are complemented by walls lined with luxurious printed leather and in the color of the champagne with which opera-goers might toast the show. To one side of the store there are niches in the wall, where hangs the brand's collection, while to the opposite there are protruding niches, rails that have been brought out from the wall yet remain enclosed in elegant frames. The shapes of the frames are octagonal and form part of a recurring yet subtle motif that extends throughout the store, from the handles on the doors into the fitting rooms, and the mirrors and hanging rails in these changing areas, to the cashier desk that appears octagonal if looked at from above. These shapes, which appear both in various forms – sometimes square, at other times elongated – add a classic yet contemporary air to the concept.

塑造商铺之王：商业店面设计 购物篇

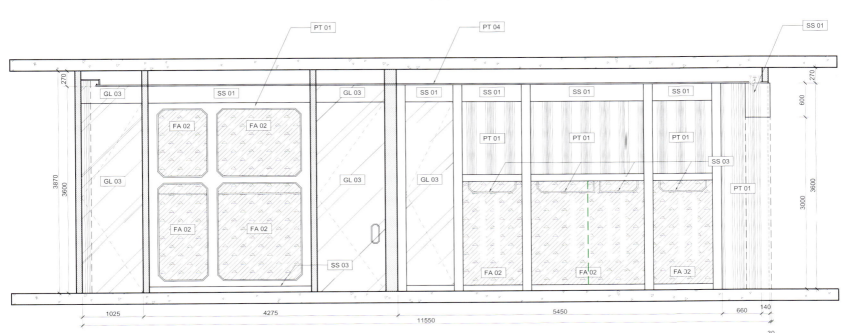

Elevation 1　立面图 1

Elevation 2　立面图 2

走进珂莱蒂尔时装专卖店，细节会成为吸引目光的焦点。顾客被店铺的波浪形墙面吸引，让人立即联想到第一幕歌戏结束后刚刚落下的帷幕。店内设有由豪华材料包裹而成的特大柔软沙发供客人休息。闪亮的石材地板由白色、灰色和珊瑚色板块组合而成。深红色的圆形地毯衬托着核心家具设计，配合椅子深紫色的色调，突显了奶黄和金色的高贵。旁边的 VIP 房专门为贵宾准备，墙边摆放着柔软的天鹅绒家具。这些家具都以豪华印花皮作为衬料，颜色以香槟色为主，让人联想到歌剧爱好者向歌唱家致敬时手中的佳酿。

店铺一边是入墙展示柜，里面挂满了品牌的独特系列，而另一边是布满精美框架造型的壁龛和栏杆。框架呈八角形，从门上的扶手延伸到试衣室，再从试衣室中的镜子及吊栏延伸到从上向下看呈八角形的收银台，形成了连续不断的精美图形。这些图形以各种形态展现，时而方形，时而长形，为整个设计增添了典雅气息，又不失现代感。

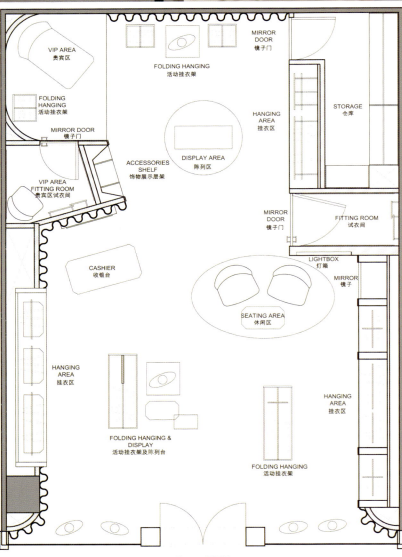

Plan 平面图

Laurèl Flagship Store Hamburg
汉堡萝丽儿旗舰店

Design agency : Plan2Plus
Designers : Ralf Peter Knobloch, Tina Aßmann
Location : Hamburg-Neuer Wall, Germany
Area : 330 m²
Photography : Ralf Peter Knobloch

设计单位：Plan2Plus
设计师：Ralf Peter Knobloch、Tina Aßmann
项目地点：德国汉堡诺伊瓦尔街
项目面积：330 平方米
摄影：拉斐尔·皮特·克诺布罗

The fashion label Laurèl presents itself as an international feminine premium brand from Germany in its new flagship store located on Neuer Wall in the main luxury shopping street of Hamburg. The overall interior design concept for this unique location was realised by the Munich architecture and interior design office Plan2Plus, who is also responsible for the international shop, showroom and headquarter concepts as well as the brand's worldwide roll out. The shop is structured by freestanding walls with jacaranda surfaces and organic fitting rooms with curtains from floor to ceiling. On the fleet side a lounge area was located in the shop's second floor that offers an unlimited view on the Alsterfleet. The Mezzanine level is used as an exclusive VIP area with integrated bar and small separated lounge area. A formally reserved, intelligent lighting concept emphasizes the sensitive design of the interior's character. It ideally validates the collections and accentuates single areas of the shop in a diverse range of lighting moods.

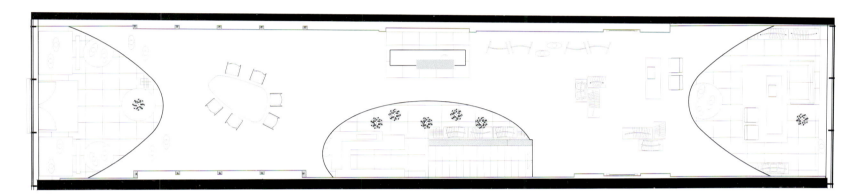

1F Plan　一楼平面图

2F Plan　二楼平面图

时尚品牌萝丽儿是一个来自德国的国际化女性品牌,其新开的旗舰店位于汉堡一条豪华的购物街上。该店的整体室内设计都是由慕尼黑的设计公司Plan2Plus完成的,这个公司不仅负责国际店面、展厅和总部的设计,也致力于品牌的全球推广。

店铺的主要组成部分是有着蓝花楹表面的独立墙以及天花板上由悬垂至地面的帘幕组成的试衣间。休憩区位于店内二层,能够将店内的景观一览无余。夹层则是一个奢华的VIP贵宾区,内设一个吧台和小小的隔离休憩区。智能照明概念强调了店内设计的特征,根据不同的灯光氛围来辨别店内的系列区和单品区。

LIU·JO Collection Milan

米兰瑠久精品女装店

Design agency : Christopher Goldman Ward Studio
Designer : Christopher G. Ward
Location : Milan, Italy

设计单位：克里斯托弗·戈德曼·沃德工作室
设计师：克里斯托弗·戈德曼·沃德
项目地点：意大利米兰

Constructive lightness, simplicity, elegance remind us of the oration philosophy underneath Ward's project. Suspension lights, armchairs and light moquette welcome the client in a home-feeling, relaxing and comfortable space which is not an issue. Space is lived and greatly enjoyed.

The project is always attractive, pleasant and coherent with the refined complexity of stings and interior settings not affecting the client, absorbed in a comfortable and never overwhelming journey. An expert eye immediately gets the smart materials choice, elements, furniture and lights making Ward's innovating project something more than a mere stylistic and intellectual exercise. Combination of style and modernity, innovation and interactiveness, a focused light system and furniture expand sensorial perception to a different experience from a simple shopping tour.

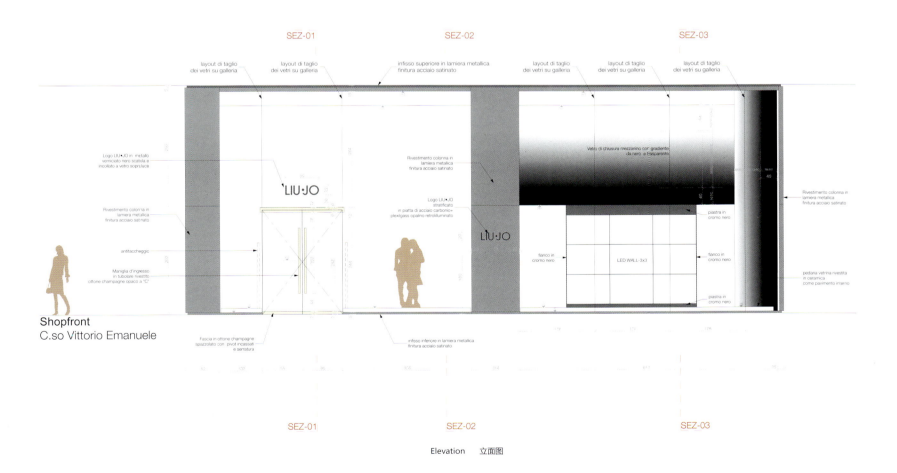

Shopfront
C.so Vittorio Emanuele

Elevation 立面图

轻盈、简洁、优雅，这就是沃德设计项目的哲学。悬挂灯、扶手椅和浅色的地毯在店内打造出一个如家一般放松和舒适的空间，整个空间都让人身心愉悦。

精致的连贯线条让整个空间富有吸引力，令人愉悦，而内部设置丝毫不影响顾客的活动，反而使其将购物变成一次舒适放松的旅行。凭借专业的眼光，你会发现材料、家具和灯光等元素都是沃德运用的极富创新性的技巧，而非风格和思维的运用。现代风格与创新和互动性相结合，以及聚焦光系统和家具，使得简单的购物拥有了不一样的体验。

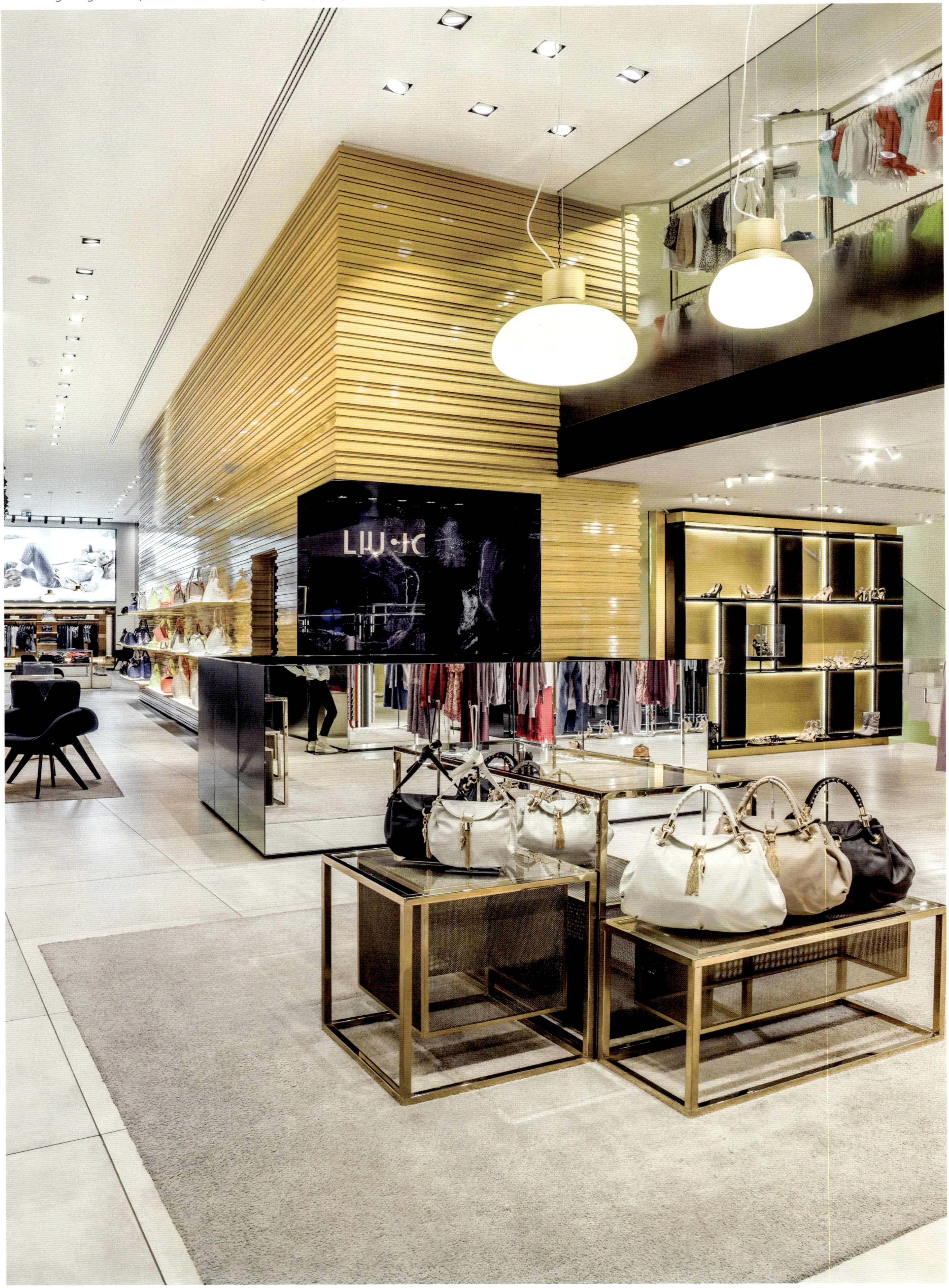
LIU•JO

塑造商铺之王：商业店面设计 购物篇

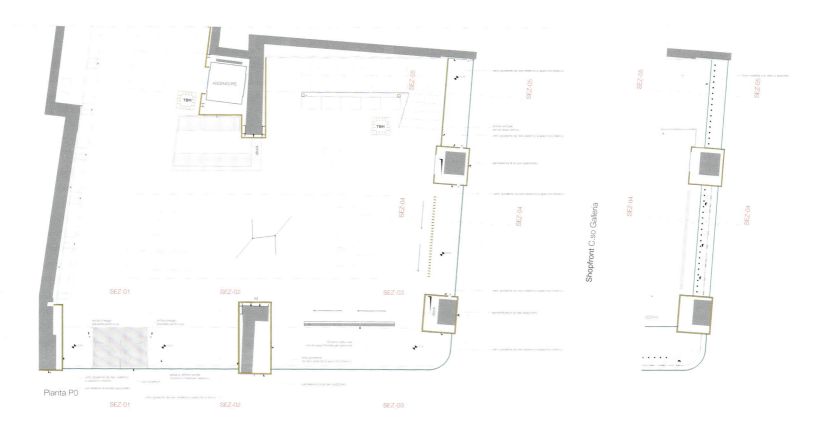

Pianta P0

Shopfront C.so Vittorio Emanuele

Shopfront C.so Galleria

Plan 平面图

Degaje Showroom
Degaje 服装展厅

Design agency: Zemberek Design Office
Designers: Başak Emrence, Şafak Emrence
Location: Merter, Istanbul, Turkiye
Area: 800 m²
Photography: Şafak Emrence

设计单位：Zemberek 设计事务所
设计师：巴塞克·艾伦斯、赛法克·艾伦斯
项目地点：土耳其伊斯坦布尔
项目面积：800 平方米
摄影：赛法克·艾伦斯

"Degaje Showroom", covers the first two floors with an area of 800 m² in the building in Merter district of Istanbul. The industrial approach of the team both at building and furniture levels, has played a role in the emergence of the concept. In both floors, suspension units, spanning the entire wall, have been designed with this approach. While water pipes and wooden shelves are used in the suspension units in the entrance floor, the suspension units in first floor were made using bent flat steel bars and plates. While the entrance floor is used entirely as a product exhibition area, in the second floor, there are locations, serving different functions. Café section is placed in a 70 m² area to allow the users to rest without leaving the location. Café is distinguished with difference in elevation, by preserving visual contact with both shopping section and meeting section.

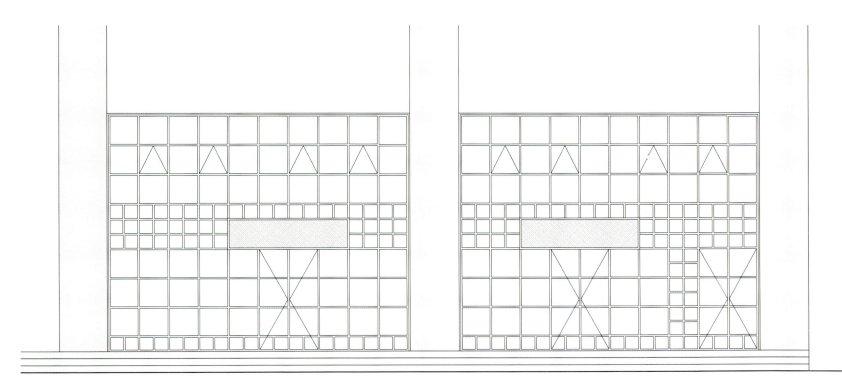

Elevation 立面图

Degaje 服装展厅位于伊斯坦布尔的梅尔泰尔区，占据了约 800 平方米的双层面积。设计团队采用具有工业气息的建筑和家具，实现了概念的设计。两个楼层都有着浓厚的工厂气息，整个墙面上都是悬挂装置。在一楼，水管和木架子都被用作悬挂装置来展示商品，楼上则采用弯曲的钢板和杆子。如此一来，一楼完全被用作产品展示区，二楼才是多功能区。内部还有一个面积达 70 平方米的小咖啡厅，可以让顾客停留在此地稍作休息。咖啡厅通过连接购物区和洽谈区，拥有较高的辨识度。

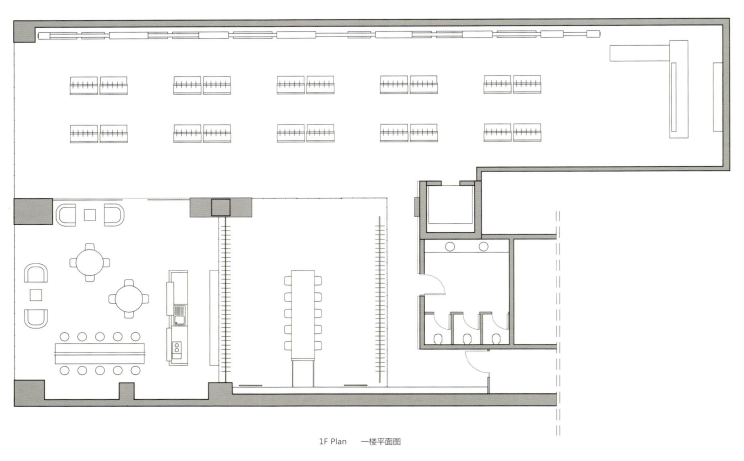

1F Plan 一楼平面图

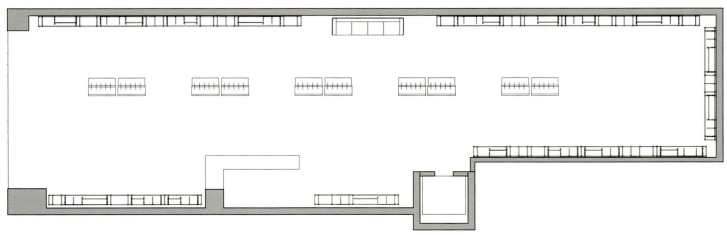

2F Plan 二楼平面图

Brooks Brothers Fashion Boutique
布克兄弟高级服装店

Design agency : Stefano Tordiglione Design Ltd
Designer : Stefano Tordiglione
Location : IFC, Hong Kong, China
Area : 80 m²
Photography : Edmon Leong

设计单位：Stefano Tordiglione 设计事务所
设计师：Stefano Tordiglione
项目地点：中国香港国际金融中心
项目面积：80 平方米
摄影： Edmon Leong

The exterior stucco plasters surrounding the columns, capitals and window are inspired by the iconic flagship, while the interior ceiling is also reminiscent of the signature building. A central cash bar greets customers as they enter the store providing a club-like atmosphere for gentlemen and women alike. Many of the store fixtures and decorations are antique, crafting a truly unique shopping environment in keep with the Brooks Brothers legacy. Furnishings in American walnut and Chicago cherry add essence to the heritage of the brand.

To showcase Brooks Brothers' as a fashion innovator with its many product introductions and high quality craftsmanship, Stefano Tordiglione Design placed large-scaled photographs zoomed-in on shirts to highlight the milestones that the brand has introduced to the fashion industry such as the first ready-to-wear clothing, the button down dress shirt and non-iron shirts.

The pale green, striped walls take their inspiration from an apartment on Park Avenue, one of the oldest and most luxurious of its kind, while on the facade, a geometric pattern is reminiscent of a classic window pane design of a 20th Century mansion situated on the Gold Coast of Long Island in New York. The motif is interspersed with the brand's prominent Golden Fleece logo. With these grand references there is a sense of luxury that Brooks Brothers has sought to reinstate in the brand.

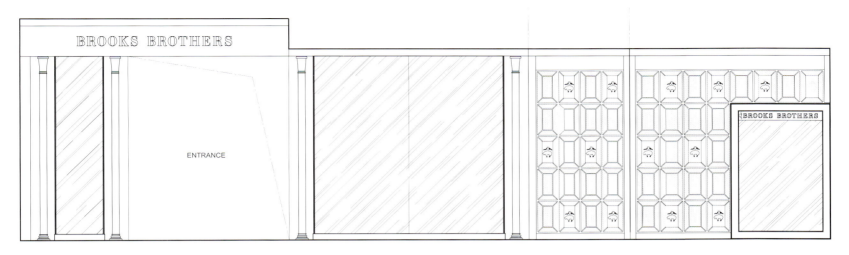

Elevation 1　立面图 1

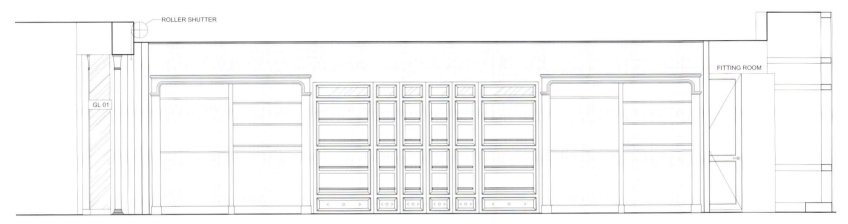

Elevation 2　立面图 2

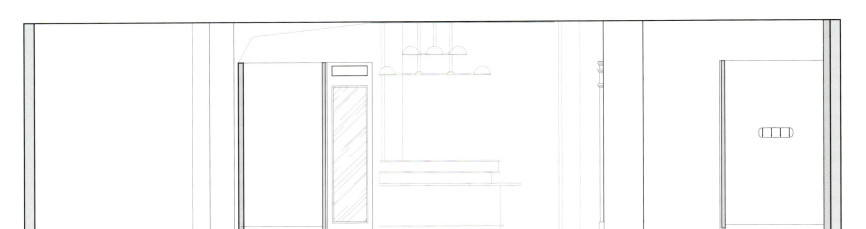

STORAGE ENTRANCE

Elevation 3　立面图 3

Elevation 4　立面图 4

店铺外墙上用以石膏为材料的圆柱、柱顶和橱窗作装饰,灵感源自纽约旗舰店的标志性设计;而店铺内天花板的装潢亦是将总部的设计引入其中。中央的吧台设计表达了对顾客的欢迎,为绅士淑女们营造了一种酒吧般的氛围。店铺内不少陈设及装饰是古董,以打造出一个真正传承布克兄弟传奇的独特购物环境;而以美国胡桃木及芝加哥樱桃木为原材料的室内陈设,更为店铺平添几分品牌魅力。

为了展示布克兄弟作为时尚潮流先驱所带来的多样化产品及高质量工艺,Stefano Tordiglione 设计事务所以大型的衬衫细节照片为装饰,暗示布克兄弟在时尚界的重要作用。扣角领衬衫,以及免熨衬衫等都一一展现于顾客面前。

以淡绿色条纹作装饰的墙壁,灵感源于纽约公园大道上一所古老而又奢华的公寓。其外墙上的几何图案设计,则来自于20世纪坐落于纽约长岛黄金海岸别墅的窗花样式。这个设计亦体现于品牌的绵羊标志之中。凡此种种,都能表明布克兄弟热切地希望让奢华感在品牌中复兴。

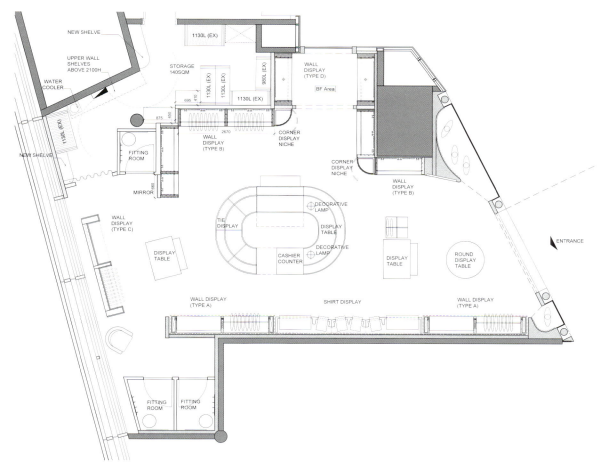

Plan 平面图

Zara Rome
罗马 Zara 概念店

Design agency : Duccio Grassi Architects
Designer : Duccio Grassi
Location : Rome, Italy
Area : 4,462 m²
Photography: Andrea Martiradonna

设计单位：Duccio Grassi 建筑事务所
设计师：Duccio Grassi
项目地点：意大利罗马
项目面积：4 462 平方米
摄影：安德里亚·马尔蒂拉东纳

At the first visit, the unexpected architectural design make the visitors perceive the original intentions of the architect: extraordinary interior volumes which open through immense windows towards the city. The former tenant had closed these windows for merchandise reasons and for technical reasons, the passage of the implants.

To restore the light, the lightness, the transparency, the thin size of the original floors became a must. The central patio cannot hosts icons as it is an icon itself. Certainly, the commercial spaces have now changed compared to the 20th century: there are needs for safety and climatization, and the amount of people has increased enormously and the merchandising is much different. The project must create the harmony between these requirements and the architecture's soul. This task is eased by the passion and the competence of the client.

On the exterior, the designers have re-opened the large windows towards the city, creating a filter, which acts as an osmotic membrane and embraces all floors inside the cube / building. Inside, each window has a double metal sheet shaped as a drapped curtain. The holes of the two metal sheets are coincident: facing them you can see outside, the sun, the rain, the city life; passing by them you perceive only the light which enters with different effects according to the time of the day.

第一次看到该建筑的时候，顾客不由得为建筑师的巧妙构思感到惊讶：令人赞叹的内部体量，通过巨大的窗户与城市对接。由于技术原因，加之商品保密原则，部分空间是封闭状态。

为了恢复室内的光线，达到一定的亮度和透明度，极有必要将原来的地板改成较薄的尺寸，中庭也成为一个彰显品牌特色的标志。当然，商业环境较20世纪也有了不小的变化：不仅有了温控和安全要求，顾客数量也大大增加了，商品也发生了很大的变化。

该项目必须在这些需求和建筑的灵魂之间找到契合点，不过得益于客户的激情和能力，这项任务的完成没有看起来那么艰难。

设计师将建筑外面与城市直接对接的窗户全部打通，再为所有楼层安装滤光镜，让光线能够射入室内。在每个窗口都设置了双层金属片，如同帷幕垂挂在窗前。金属片板上面都有开洞，当你从里往外看，能够看到阳光雨滴，将城市景色尽收眼底；而从外往里看，则能看到店内随着时间推移而不断变化的灯光效果。

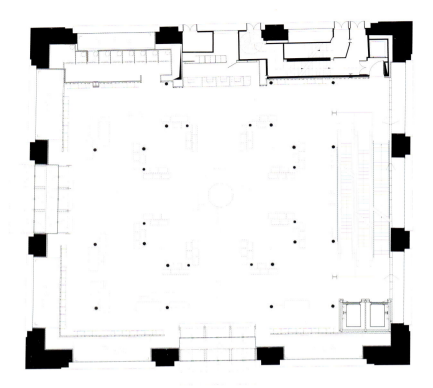

1F Plan 一层平面图

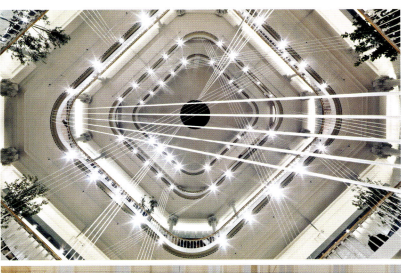

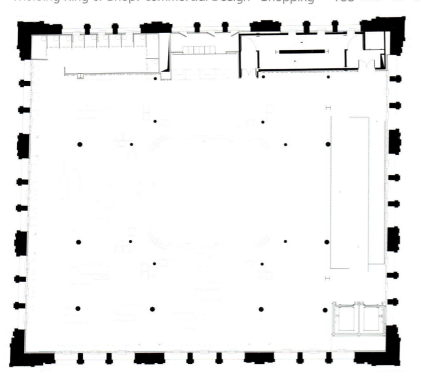
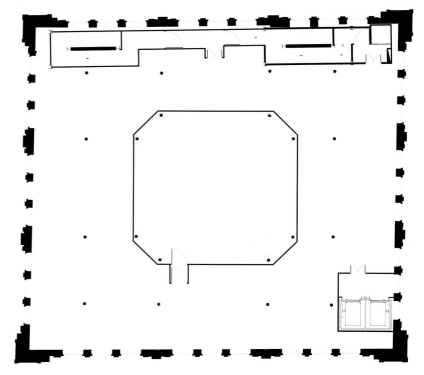

2F Plan　二层平面图　　　　　　　　　　　　　　　　　　　3F Plan　三层平面图

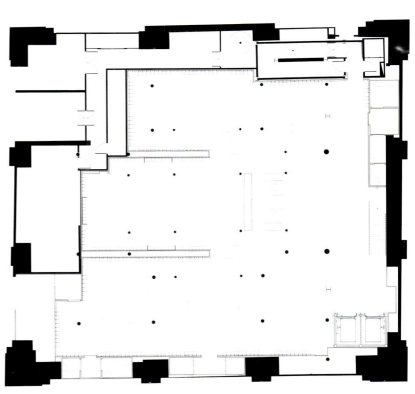

Basement Plan 地下室平面图

Classic Precipitated Luxury – Giuseppe Custom Professional Service Center

经典沉淀的奢华——乔治白职业服高端定制

Design agency : Shanghai Window Design
Designers : Li Xiu'er, Hu Wanquan
Location : Beijing, China
Area : 210 m²

设计单位：上海尚窗室内设计有限公司
设计师：李秀儿、胡万权
项目地点：中国北京
项目面积：210 平方米

This is the first listed enterprise of business wearing; Giuseppe Custom sets a service center here to showcase the materials and manufactures as well as display the ready-to-wears. With a VIP area for body measurement, this case offers a much nobler shopping experience, thus interpreting its background culture and brand value. Inheriting the traditional classic of European-custom stores, Giuseppe stresses on the competent and commercial qualities which are concise and elegant. Also, it pays attention to the textures of the original materials, a sense of luxury through the material and details of manufacture.

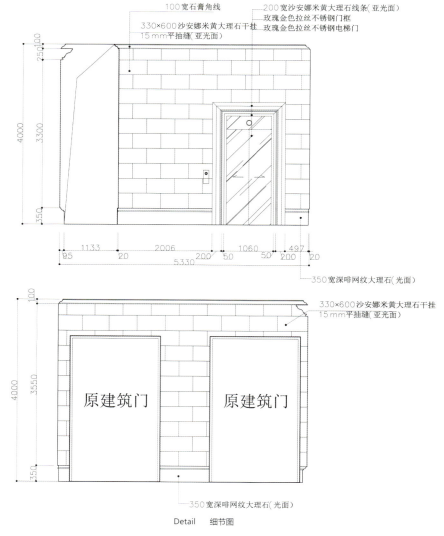

Detail 细节图

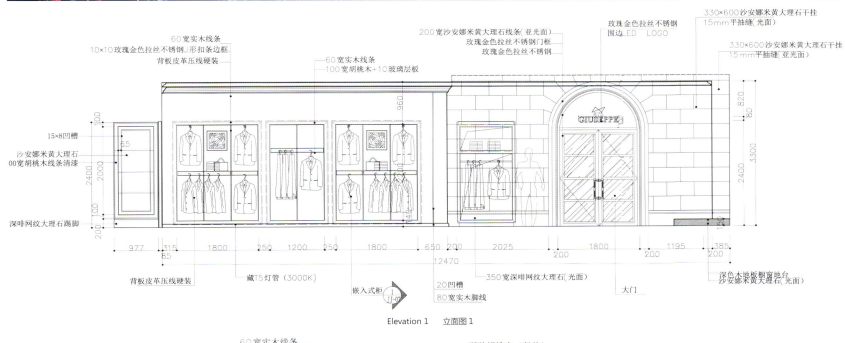

Elevation 1　立面图 1

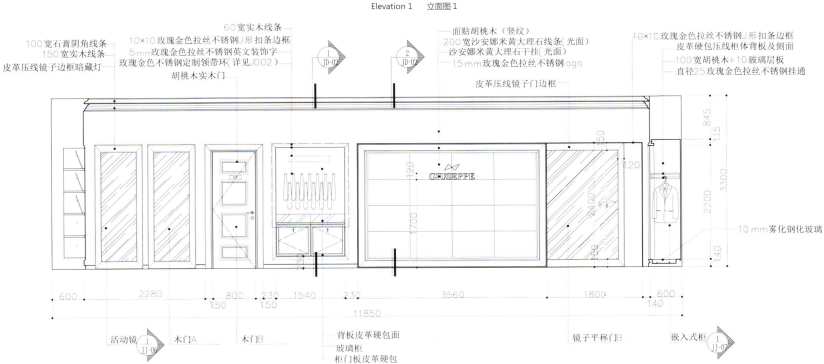

Elevation 2　立面图 2

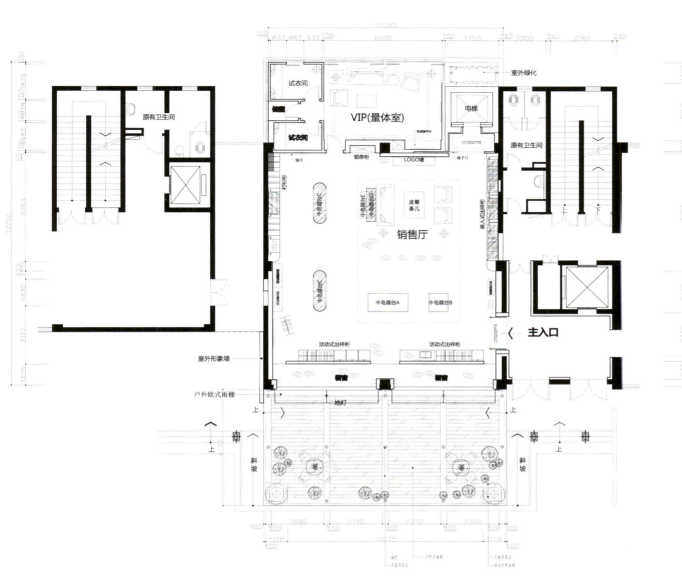

Plan 平面图

该案业主创建了中国第一家职业装上市企业。该案是为乔治白职业服高端定制系列而设立的销售中心，除展示成衣外，还有面辅料与制作工艺的展示，并设有供量体使用的VIP区。该案诠释了乔治白品牌的背景文化和品牌价值，同时为顾客提供了更尊贵的购物体验。乔治白品牌传承传统欧洲定制店沉淀下来的经典，强调职业装的干练与商务感、简约与华丽感，注重材质本身的质感，通过材质与工艺细节彰显奢华。

BUBIES Lingerie Flagship
BUBIES 内衣旗舰店

Design agency : PplusP Designers Ltd
Designer : Wesley Liu
Location : Hong Kong, China

设计单位：维斯林室内建筑设计有限公司
设计师：廖奕权
项目地点：中国香港

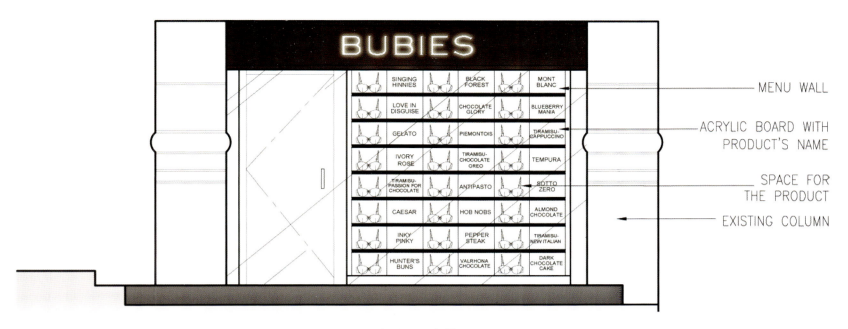

Elevation 1　立面图 1

BUBIES Lingerie Limited is a newly launched brand which aims at redefining the meaning of lingerie. The materials used in this store are truly sustainable, such as LEED accredited wall papers, zero toxic emission paints and LED reflector lights. The first illustration begins with the gigantic display wall which serves as the window display; it is also like a large menu that contains appetizers, main courses and desserts. The shop interior is themed with massive use of glass boxes and bird cages in different sizes. It is believed that ladies' underwears are something to protect and show off their graceful bodies, like a beautiful bird being kept in a bird cage showing off its beauties.

Moreover, the products are being displayed in a very unique way. As a coherent of the menu display wall, the products are displayed in fine porcelain and glass tableware; some products even look like a cupcake, which also means that their lingerie products are as fine as or as detailed as delicious premium desserts.

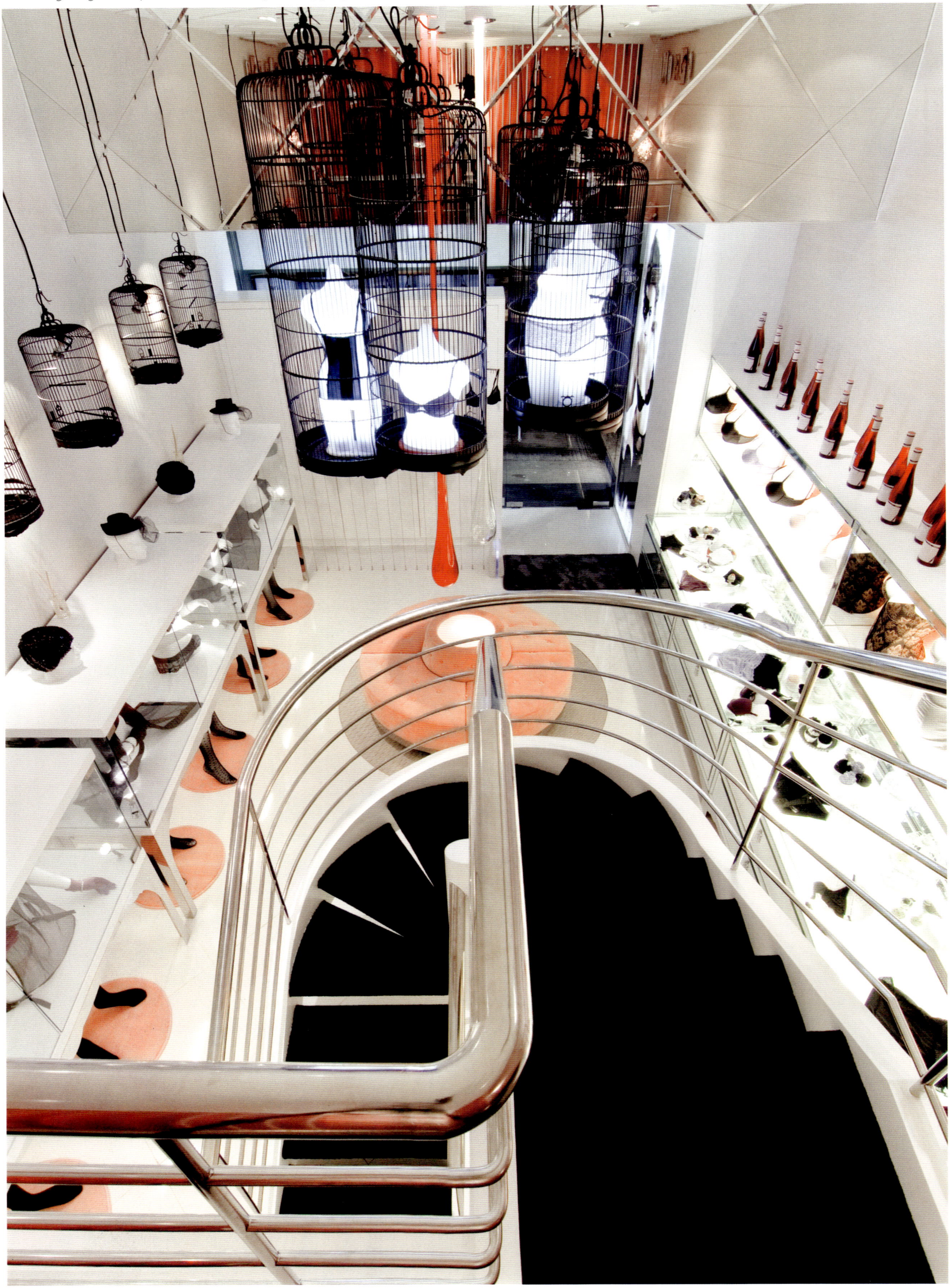

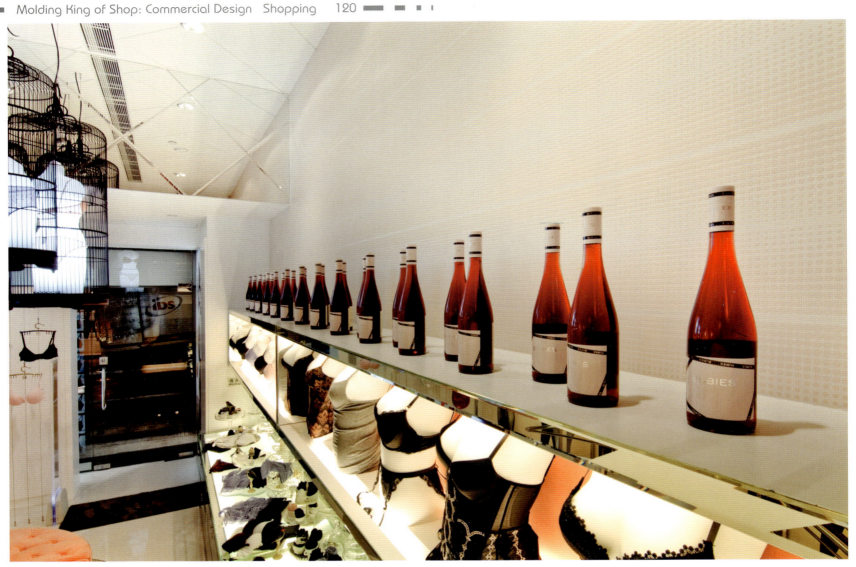

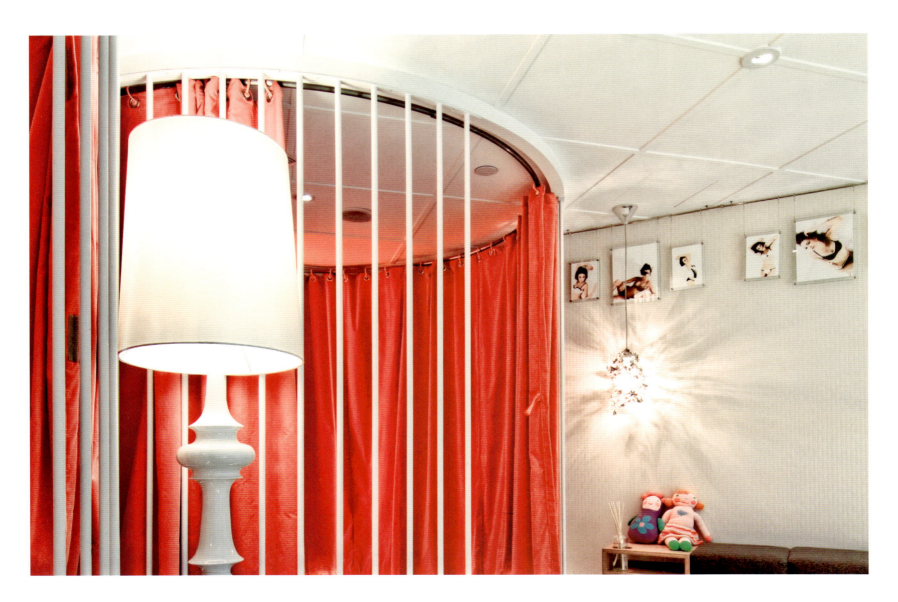

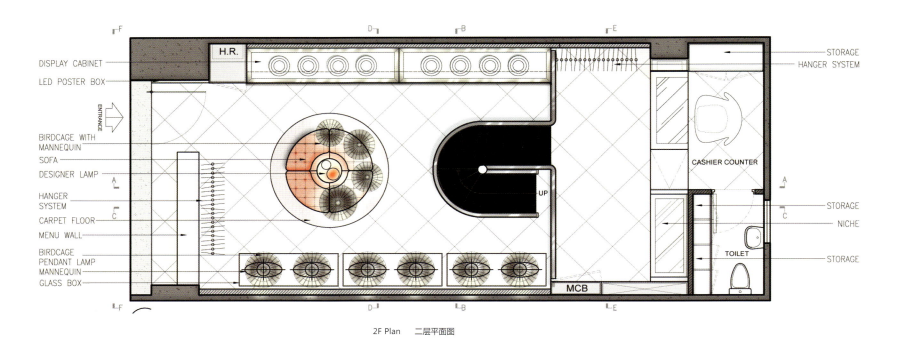

1F Plan 一层平面图

2F Plan 二层平面图

Elevation 2 立面图2

Elevation 3 立面图3

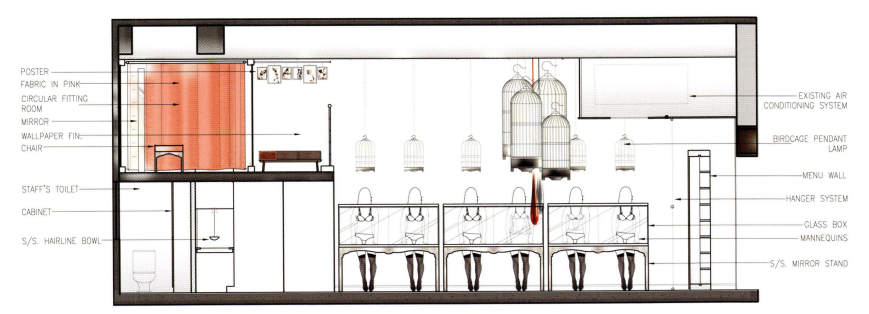

Elevation 4　立面图 4

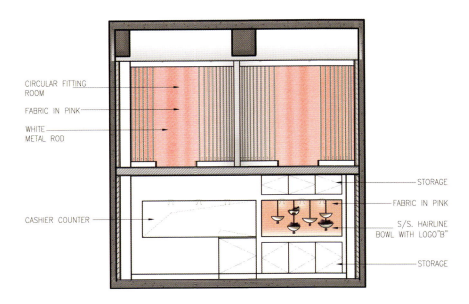

Elevation 5　立面图 5

这是一个女性内衣专卖店。全店均配置了 LED 反射灯，在节能的同时亦营造出璀璨夺目的灯光效果，而墙纸、饰品用料均选择不含有害物质，并且符合 LEED 国际环保标准的产品。由橱窗开始，大型展示柜犹如一份菜单，店铺内的"菜色"顿时一览无遗，在展览的同时亦与主题相呼应，不同的内衣化身为一道道精致的甜品。顾客亦会发现店内挂满了鸟笼装饰，就连阁楼的更衣室亦如鸟笼一般，让人联想起在鸟笼里花枝招展的孔雀，内衣如鸟笼一般既可保护亦可展现女性的优美身段，与品牌的理念非常契合。

进入店铺后，长桌上的银制餐盘、陶瓷碟子、玻璃瓶内均摆放了不同的内衣商品，商品犹如高级甜品般精巧、优质，顾客可以如享受自助餐般无拘无束地挑选称心如意的商品。

Elevation 6　立面图 6

Ecko Unltd
犀牛服饰

Design agency: Stone Designs
Location: Madrid, Spain
Area: 80 m²
Photography: Stone Designs

设计单位：Stone 设计事务所
项目地点：西班牙马德里
项目面积：80 平方米
摄影：Stone 设计事务所

Ecko is a brand, which cannot be easily measured or classified. Ecko represents the most rebellious heart of NYC; it is the response to the adoption of bourgeois fashion, and the commitment to the worldwide fashionable trend. Ecko deserved a store image that has the feeling of authentic NYC from the 1980's and 1990's, those years in which NY was a high-spirited city in which the most exciting social, artistic and cultural events have taken place. The designers' idea is to mix the best things from the past with the present New York icons that will last. The designers want to show clearly how a city and a culture may have influenced our lives and how Ecko is a brand that has not only been influenced by NY, but also been part of it, contributing to NY and continuing to do so actively to keep NY alive.

Just like a good piece of art, a good project is solved with few elements that express themselves well. The store is a space where customers will perceive simply what Ecko represents and feel part of it, knowing that they are important to the brand. The sources of inspiration have been urban icons, the simple elements like scaffolding, which is a structure that has covered and will always cover the streets of Manhattan. Other elements are the construction fences, made in OSB boards and displayed in a random order and painted in white to convey a warm texture, putting out of context, a material that will provide a new quality. This store is the mixture of the three most representative elements of the New York life and culture, remodeled to make people feel the purest essence of Ecko and that when they are inside the store, they are in the iconic city—NY.

Elevation 1　立面图 1

"犀牛"是一个时装品牌,它不能被简单地评介或归类。犀牛服饰象征着"纽约最叛逆的心",也是对资产阶级时尚的回应,更是这个城市对世界时尚潮流的承诺。犀牛服饰的形象代表是20世纪八九十年代的纽约,那些年,纽约是一个意气风发的城市,是最激动人心的社会、艺术和文化大事件的发生地。该项目的设计理念就是将纽约过去的和现在的流行元素进行混搭,从而清晰地呈现一座城市和一种文化是如何影响人们的生活。犀牛服饰受到纽约文化的影响,为纽约的时尚潮流助力,并体现出鲜明的品牌特色。

就像一件好的艺术品一样,一个优秀的项目也需要一些设计元素将本身的魅力展现出来。店铺是一个能够让顾客轻松体会犀牛服饰的内涵的空间,并让他们觉得自己是其中的一分子。设计灵感来源于城市标志,以及一些简单的元素,如脚手架———一种覆盖了曼哈顿所有街道的结构。其他一些如建筑围墙(由欧松板制作而成)被涂成白色并随机摆放,营造出温馨的氛围,其脱离了整体环境,也能够创造出新的品质。这个店铺是纽约生活和文化中三个最具代表性的元素的整合,让人们感受到犀牛服饰最纯粹的本质。走进商店,仿佛置身于时尚艺术气息浓郁的纽约市。

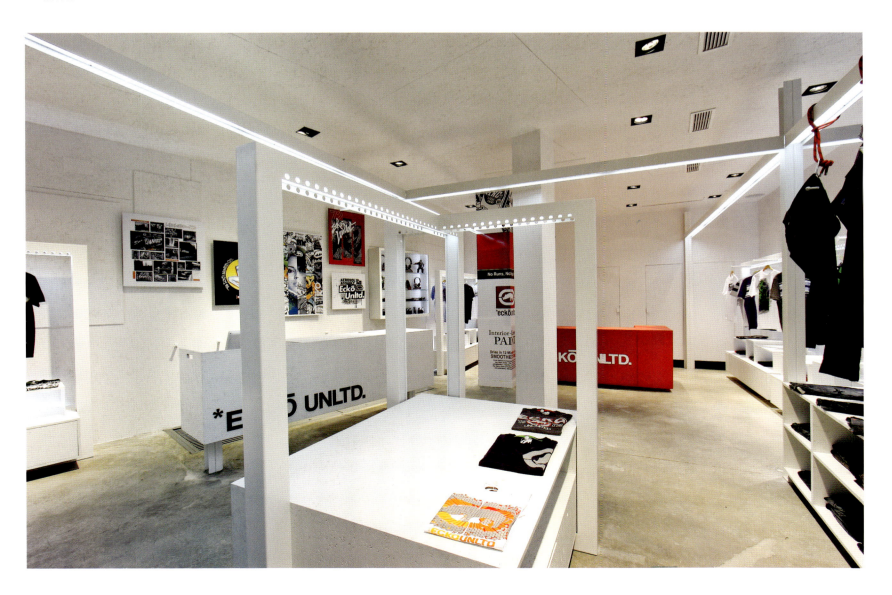

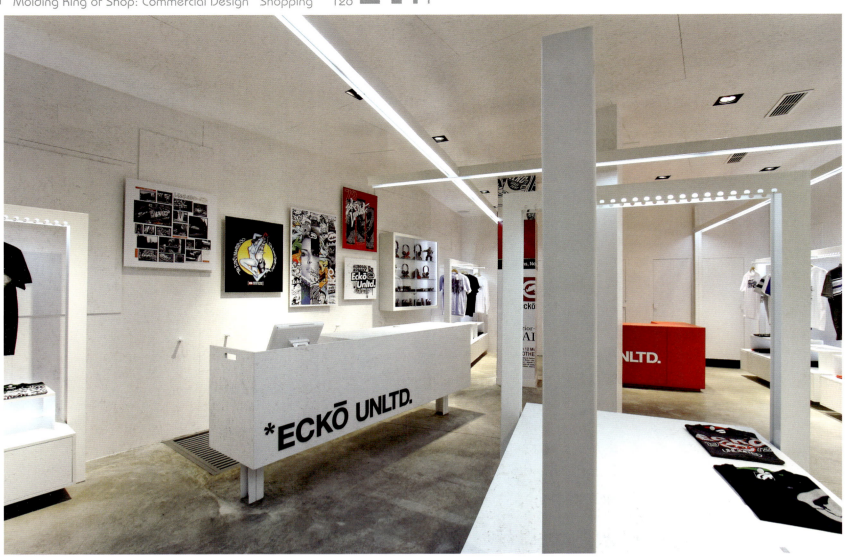

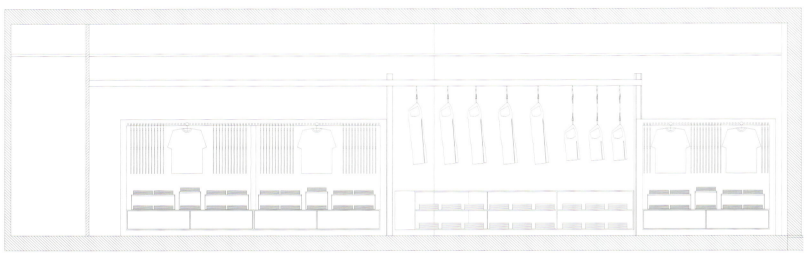

Elevation 2 立面图 2

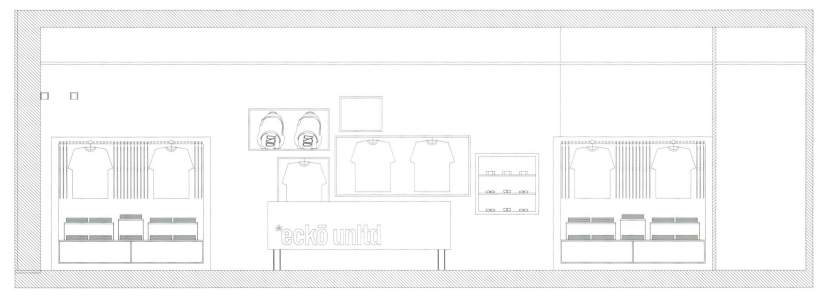

Elevation 3 立面图 3

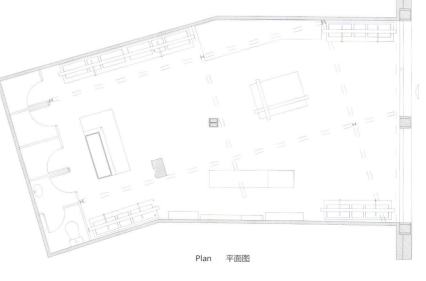

Plan 平面图

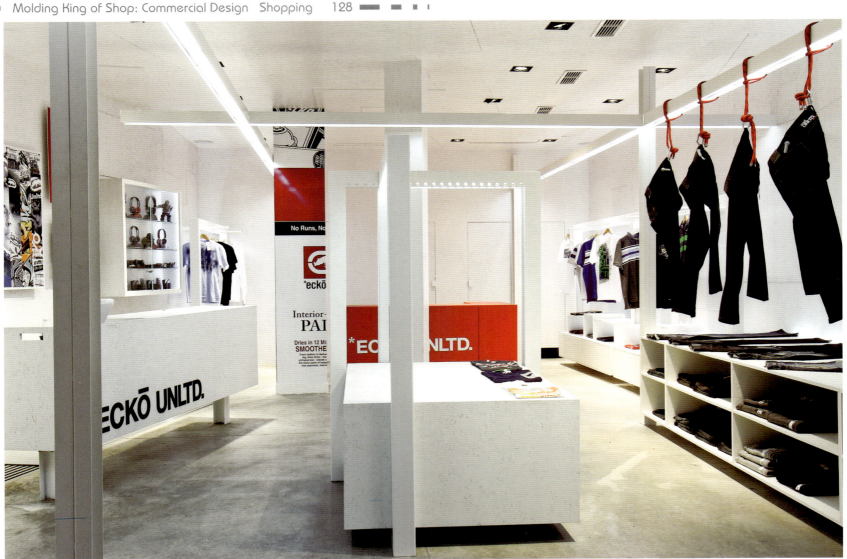

Hanna Trachten Flagship Store in Vienna

Hanna Trachten 维也纳旗舰店

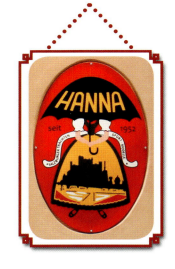

Design agency : 5 Star Plus Retail Design
Designers : Barbara Seidelmann, Bruce Li
Location : Vienna, Austria
Area : 110 m²
Photography : Barbara Seidelmann

设计单位：斐思达品牌设计咨询有限公司
设计师：Barbara Seidelmann、Bruce Li
项目地点：奥地利维也纳
项目面积：110 平方米
摄影：Barbara Seidelmann

The brand HANNA Trachten, which was founded more than 60 years ago, is positioned as a provider of traditional clothing with a modern twist. For the Vienna flagship store located in the capital's central shopping district, the design team created a themed, yet elegant store design scheme which features some traditional Austrian furniture pieces. The farther customers walk into the store, the more it resembles a fairy tale world: Customers transit from the city into an enchanted forest.

The design in the first rooms plays with the vaults of the medieval building and references Salzburg. A virtual passage into a mystic world, a corridor which leads to other sales rooms and displays children's clothing, with features illustrating animals of the forest on the walls. All shelving fixtures of this store design project were custom-designed to fit the vaulted and irregular walls.

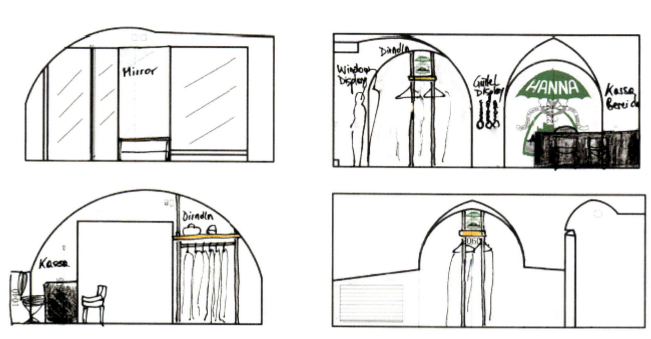

Elevation 1　立面图1

To promote brand awareness, a strong focus was kept on positioning branding elements throughout the store. The waiting room was transformed into a branded environment which was a perfect backdrop for the propagandas for lamp box and brand introduction on display.

Hanna Trachten 品牌拥有超过 60 年的历史，是奥地利传统服装与现代风格相结合的成功代表，所有服饰均由自己设计、制作和销售。该旗舰店位于维也纳最繁华的商业中心区域，设计团队为其设计了主题特色鲜明、极具优雅感的店面，再搭配奥地利传统家具，使顾客置身其中，能够充分感受如童话世界般的购物氛围，体验从城市生活到魔幻森林的神奇转变。

商店内的第一个购物区是在中世纪城镇建筑的拱顶的基础上进行设计的，并借鉴了其品牌发源地萨尔斯堡的城市建筑风格。而用来连接不同销售区的走廊，好像一个通往神秘世界的通道，用来展示儿童服装，并在墙上用森林中的小动物作为装饰点缀。商品货架均是设计师为这里量身定做的，以便与空间本身的拱顶和弧形边墙完美搭配。

为了提高品牌认知度，贯穿于整个商店的品牌元素都被重点突出设计，包括品牌标志、形象物等都被放置于墙和货架的醒目位置，而灯箱宣传画和品牌背景介绍被设置在顾客等候区内。

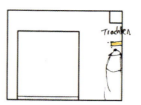
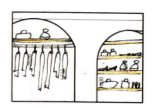
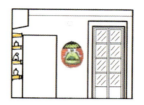

Elevation 2　立面图 2

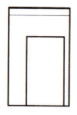

Elevat on 3　立面图 3

House Layout Drawing 户型图

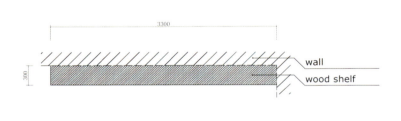

Top View

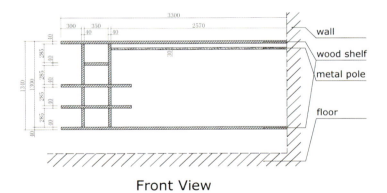

Front View

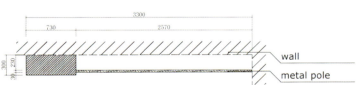

Left View

Elevation 4 立面图 4

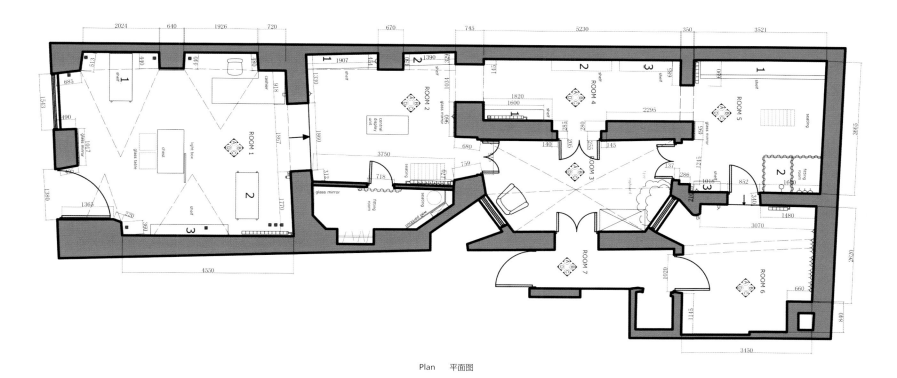

Plan 平面图

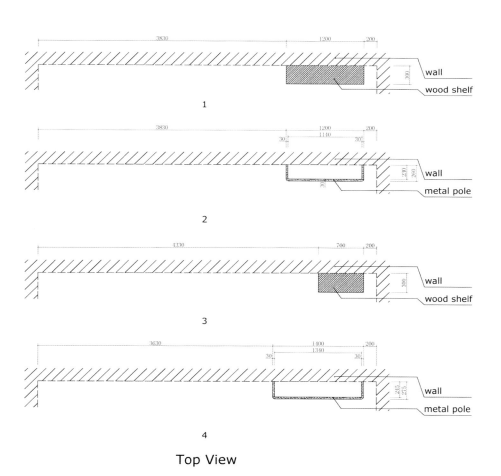

Top View

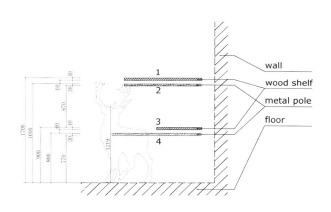

Front View

Left View

Elevation 5 立面图 5

MASH Flagship Store in Beijing
MASH 北京旗舰店

Design agency : 5 Star Plus Retail Design
Designers : Barbara Seidelmann, Bruce Li, Ray Zhang
Location : Beijing, China
Area : 116 m²
Photography : Bruce Li

设计单位：斐思达品牌设计咨询有限公司
设计师：Barbara Seidelmann, Bruce Li, Ray Zhang
项目地点：中国北京
项目面积：116 平方米
摄影：Bruce Li

MASH, a stylish European fashion brand from Verona, Italy, is positioned in the upper medium segment of the market. The collections are casual yet chic, with Italian sophistication apparent in the quality of materials, especially the detail of the garments and the overall design of the couture. This approach had to be reflected in the atmosphere of the store.

For the brand's flagship store in Beijing's Solana Mall, the design team created a store design inspired by a conceptual design scheme provided by the Italian brand as well as the geographic environment of the fashion producer – the city of Verona. The main design goal was to use the store design to tell the brand's story and to create an easily recognizable connection to the city of Verona, in order to clearly distinguish MASH from its competitors in the local Chinese market.

The display areas for MASH Queen and MASH Uomo, the women's and men's fashion collections, are created as islands to the right and left side of the store with different flooring and wall finishes. While the space for MASH Queen has a more sophisticated feel thanks to dynamic display fixtures with slightly varying shelf heights and bright mosaic flooring, the area around Mash Uomo is more masculine, characterized by a used wooden floor and a 3 meters high custom-made casual display unit. The central floor space is kept in a neutral grey shade and is a reference to Verona's main river Adige, drawing customers to the store's back wall featuring arcades with abstract black-and-white views of the city. One of the arcades is open and leads to the fitting rooms.

MASH，一个来自于意大利名城维罗纳的欧洲时装品牌，它定位于中高端市场，其服装兼具高雅及休闲的风格，特别是在某些细节处理及整体设计方面，将意大利精致的艺术设计理念体现在高品质的服装面料上。同样，这样的设计思路也需要在对其零售店的整体设计理念中加以体现。

对于这家位于北京蓝色港湾的 MASH 品牌旗舰店，设计团队通过对品牌发源地维罗纳的历史背景、地域环境特点等相关信息进行分析，完成了整套店面设计的理念构思。主要的设计目标是运用数种设计元素的组合来诠释这个品牌独特的历史渊源，并搭建一条与其发源地——维罗纳紧密相关的联络线，从而使 MASH 的品牌形象从中国市场中脱颖而出。

商品展示区中分隔出的 MASH Queen 和 MASH Uomo，即女装区和男装区，通过运用不同的地板和墙面设计，分别被安置在商店两端的展示岛上。女装区中，轻便简洁的货品展示架和明亮的马赛克地面的混合搭配彰显出女性柔美的气质，男装区则通过实木地板与 3 米高的特别定制的商品展示架的搭配，彰显出男性阳刚的一面。二者之间的地面则使用了维罗纳城市的形象象征——Adige 河的中性灰色，将顾客的注意力吸引到商店的拱廊墙上，那里以黑白风格展现了维罗纳的城市风光。其中一个拱廊是开放式的，可以通向试衣间。

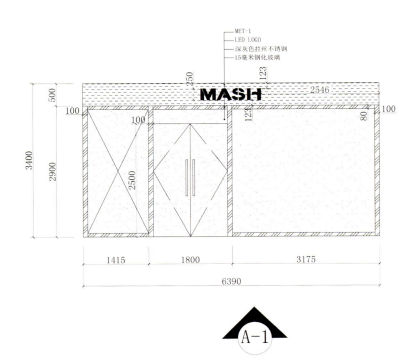
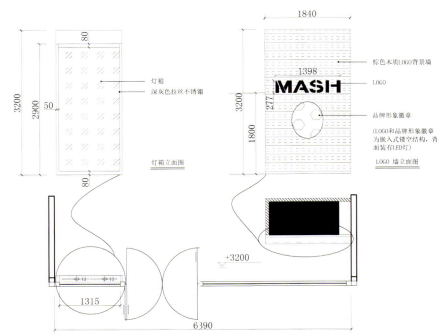

Elevation 1　立面图 1

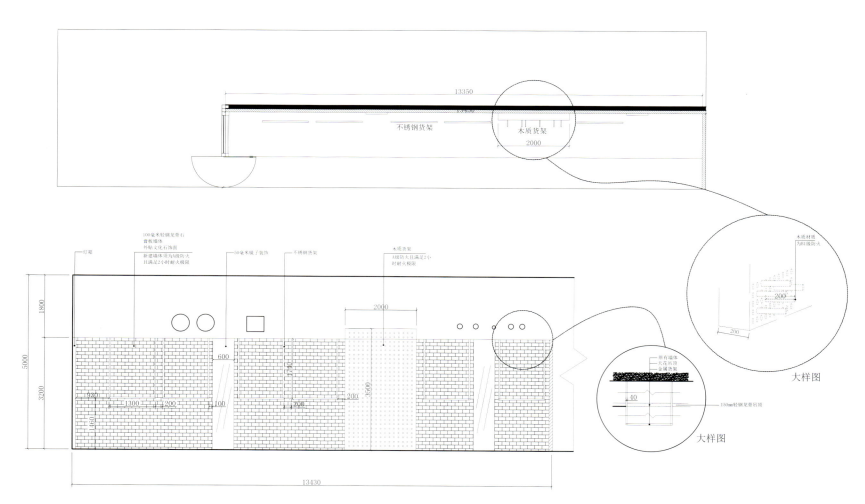

Elevation 2　立面图 2

Molding King of Shop: Commercial Design Shopping 140

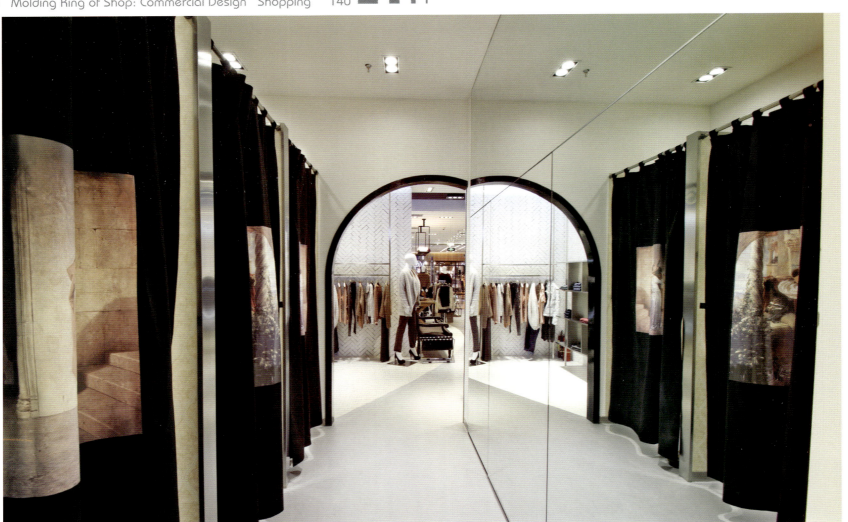

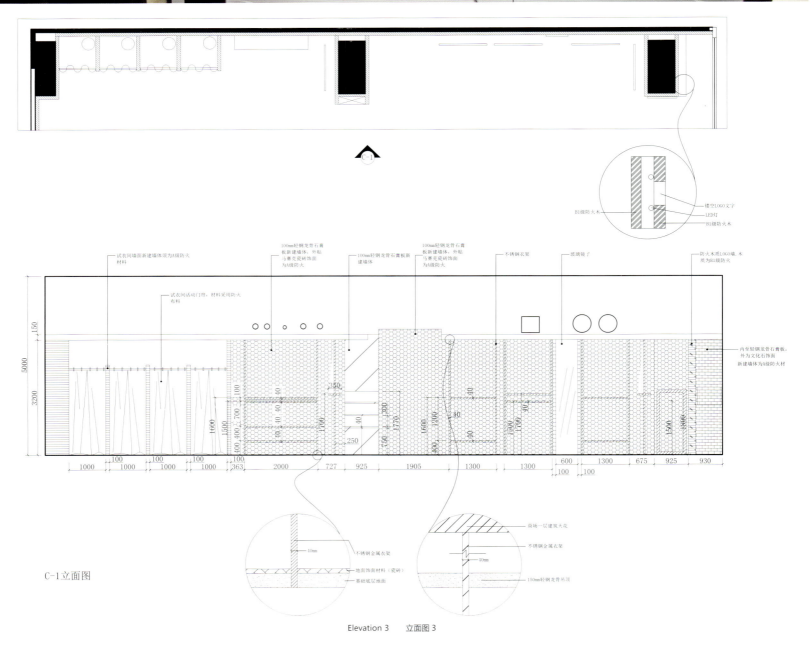

C-1 立面图

Elevation 3　立面图 3

Elevation 4　立面图 4

MUJI Shop in Milan
米兰无印良品

Design agency : Roberto Murgia Architect,
　　　　　　　Aliverti Samsa Architect
Designer : Miriam Cesco
Location : Milan, Italy
Area : 250 m²
Photography : Giovanna Silva

设计单位：罗伯特·莫吉亚建筑事务所、
　　　　　艾丽弗缇·萨姆沙建筑事务所
设计师：米里亚姆·赛思科
项目地点：意大利米兰
项目面积：250 平方米
摄影：乔凡娜·席尔瓦

The new Muji Store is located in Gae Aulenti Square, a new public space near Garibaldi Station in Milan, also in the heart area for EXPO 2015.
The shop is a double-height space of 250 m² with a 35 meter long suspended walkway. The philosophy of the brand is reflected in the design of the shop; the keyword is essential. Every Muji Store is made to exhibit the products; it's a showcase without showing off. The material used are natural: wood, brick, iron and plaster; they are in perfect agreement with the simple Muji products. Natural wood shelves are used to accommodate merchandise.
The same finish is for parts of the wall; the other parts are brick walls with original bricks from Asia. The upper side of the wall includes graphics with Muji products designed by Kenya Hara. The iron walkway is white painted; it hangs from the ceiling with thin rods. Parapet is perforated with thousands of squares of different size, these make the heavy structure a permeable object. The shop has large windows and the natural light fills the space. The entrance window is on the square at the second level, while the opposite window is on the street, at the third level. The white perforated stair looks like a soft frame.

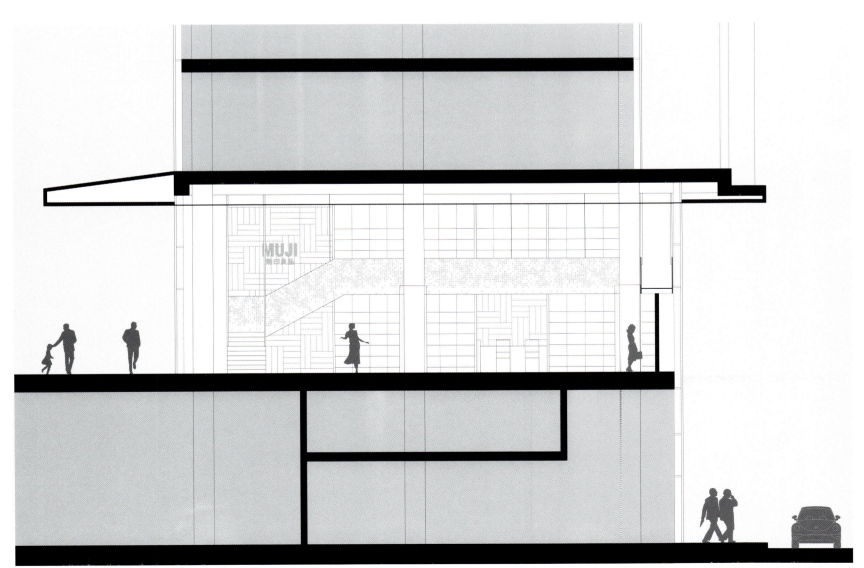

Elevation 立面图

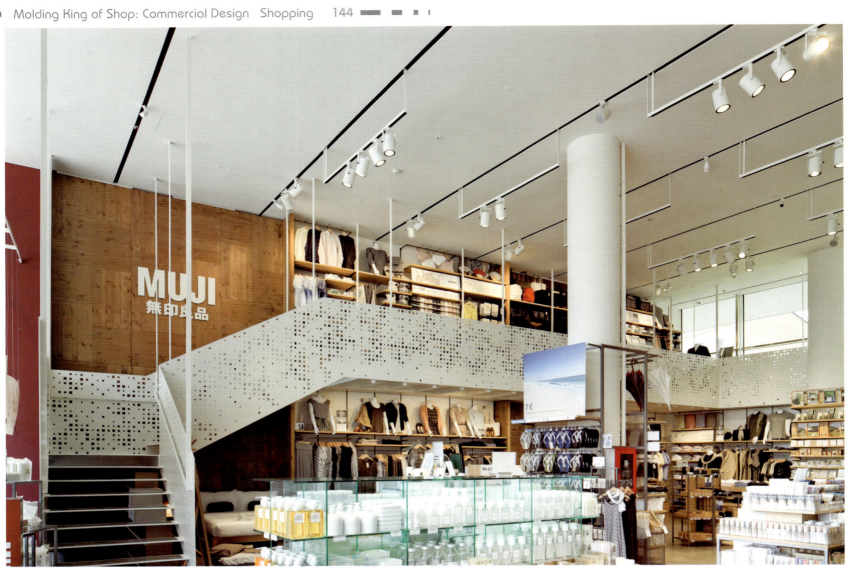

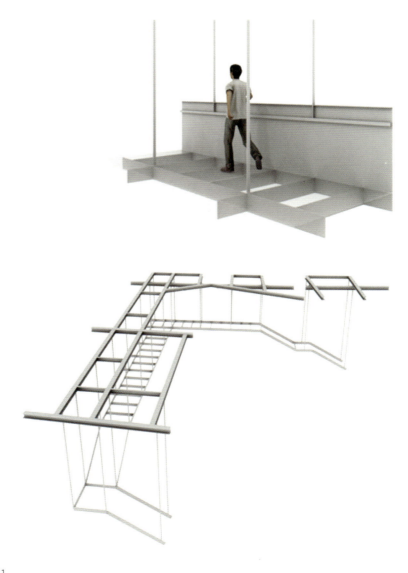

Detail 1 细节图 1

Detail 2 细节图 2

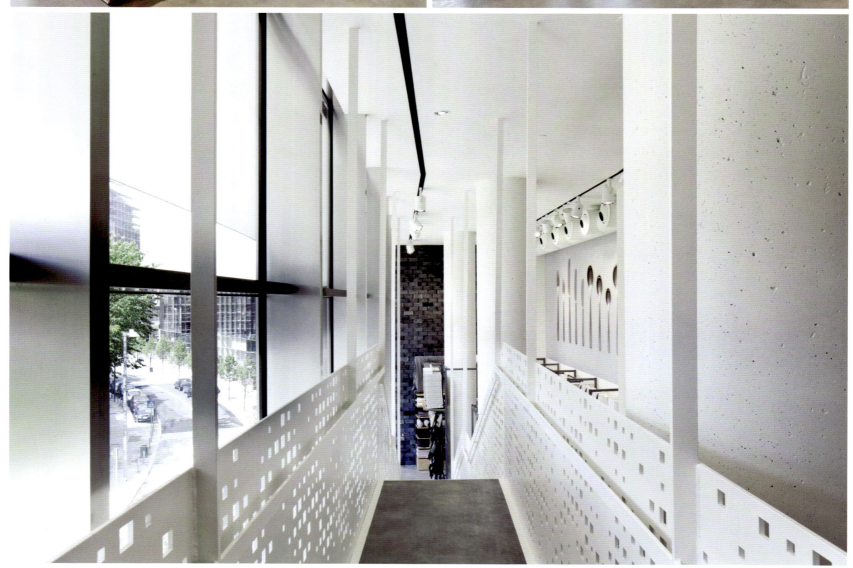

MUJI 新店面由意大利 Roberto Murgia Architect 与 Aliverti Samsa Architect 合作完成，它位于米兰 Gae Aulenti 广场，一个靠近 Garibaldi 车站的新公共场所，同时这里也是 2015 年世博会规划区域的核心位置。

新店面是一个 250 平方米的双层挑高空间，并拥有一个 35 米长的空中走廊。无印良品的品牌哲学被设计师很好地反映在店面设计中，其中最为关键的是每一个无印良品的商店都要以简练的陈列形式突出其所展示的商品。材质首选自然的木材、砖、铁和涂料，以契合商品的自然属性。

除了统一的白色涂料和木板墙之外，设计师在靠近外部空间的墙面上铺设了来自亚洲的深灰色砖。被完全漆成白色的空中走廊吊装在天花板上。走廊扶手上密布着大小不一的方形开口，这些冲孔赋予厚重的结构以半透质感。店面一侧巨大的窗户让自然光溢满空间，另一临街立面的窗户则在二层高度，从外看去，白色冲孔楼梯精致且柔美。

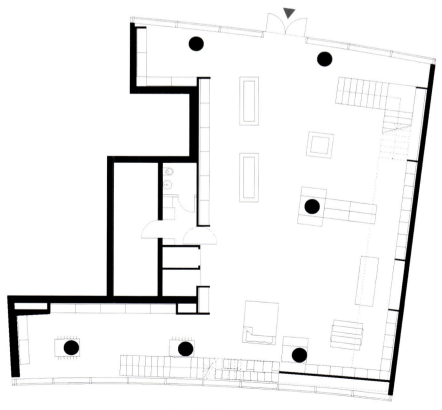

Plan 平面图

Ora Creation Store
Ora Creation 概念店

Design agency: 5 Star Plus Retail Design
Designer: Bruce Li, Barbara Seidelmann, Ray Zhang
Location: Beijing, China
Area: 85 m²
Photography: Bruce Li

设计单位：斐思达品牌设计咨询有限公司
设计师：Bruce Li, Barbara Seidelmann, Ray Zhang
项目地点：中国北京
项目面积：85 平方米
摄影：Bruce Li

The new Ora Store in Beijing's Solana mall was to be a multi-brand store selling a number of European niche brands, which are not available anywhere else in China. The brief required an elegant, playful and neat store design where the brand could express its personality and provide customers with a design-oriented, yet dynamic European lifestyle. Ora now has a fully-designed retail store in Solana mall in Beijing that successfully reflects their sophisticated, yet modernistic brand.

Ora, like its unique fashion, needed to stand out among its surroundings. Asking for a flexible space with a dedicated focal area for brands, Ora wanted a warm color scheme and dynamic concept with some energy and contrast. Obtaining materials that were available at a limited cost were flexible enough to be sculpted into dynamic shapes and that added some structural strength was a task in itself. Inspiration for such materials was taken from existing designs in the nature, such as spider webs, fishing nets, woven elements and metal mesh used for fencing. The overall design features a black shiny entrance with "ORA

Elevation 1　立面图 1

Creation" illuminating above in LED light to give a bold contrast to the more neutral shades behind the logo. On the left of the entrance is a diagonal, vertical wall where a panel is fixed, where various brands from different designers are sold there.

Enticing customers from afar, the wall decoration was the key design element of the store. The effect was actually created with two layers—a grey wall with sporadically flowing shapes painted in red and gold, and in the front is the second layer made of a natural white metal mesh. This creates a sense of depth and movement when customers walk through the store. To draw customers to the back, the store design implemented an accessory display shelf and a curved counter painted in a vibrant red. The mirror wall and fitting room paralleled with each other offer a balance between modernity and traditionalism, once more underlining the unique luxuriousness of Ora.

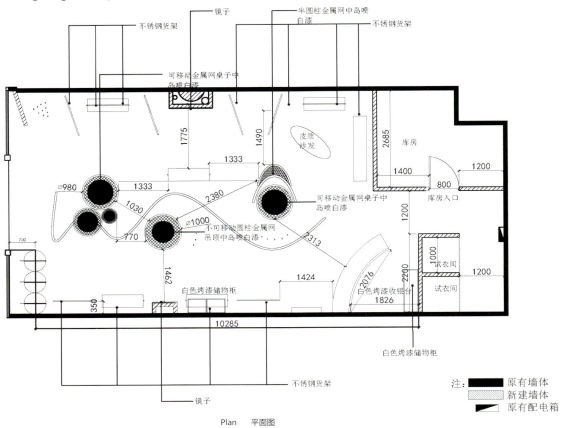

Plan 平面图

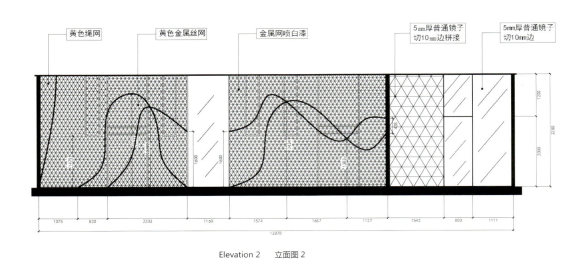

Elevation 2 立面图 2

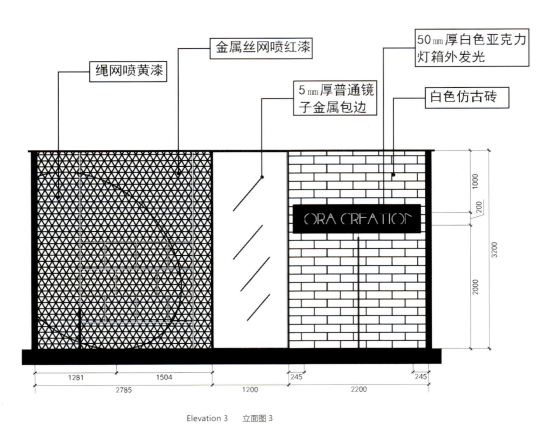

Elevation 3 立面图 3

Molding King of Shop: Commercial Design Shopping 152

- ① 圆柱展架
- ② 金属原有展架
- ③ 白色烤漆展柜
- ④ 白色烤漆收银台
- ⑤ 储藏室
- ⑥ 试衣间
- ⑦ 拼镜
- ⑧ LOGO 墙

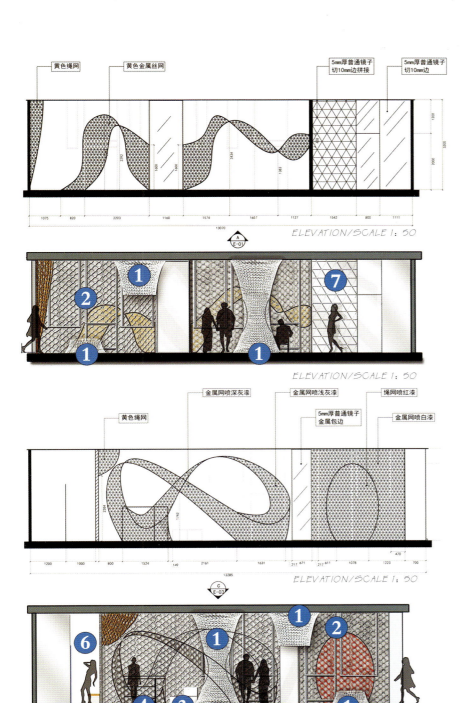

新的 Ora 店位于北京的蓝色港湾购物中心，是一个多品牌商店，主营在中国其他城市无法购买的欧洲服装设计师品牌。业主需要一个优雅、俏皮、利落的店面设计，以宣扬品牌的独特个性，并为顾客提供具有设计感及充满活力的欧洲生活方式体验。这个风格新颖的零售店，极具成熟而独特的现代品牌设计风范。

Ora，就像其独特的品牌风格一样，需要从其商店周围的环境中脱颖而出。空间设计需要考虑品牌空间的活动性，采用了一个温暖的颜色主题和一些动感和强对比度的概念方案。设计采用低成本的材料，具有足够的灵活性，结构强度适中。材料的设计灵感来自于大自然，如蜘蛛网、渔网、用于围栏的编织件和金属网。品牌标志"ORA CREATION"被安置在黑色入口的上方，配以 LED 灯，以便和背景的中性色形成一个大胆的高对比度反差。入口的左侧是一面垂直的墙，购物架与之固定在一起，这里出售不同设计师的品牌服装。

为了吸引从远处经过的顾客，墙面装饰是商店设计的关键元素。整体效果按照形状被分成两层：一层是配有红色和金色流动图案的灰色装饰墙，另外一层是在其前面专门设计的白色金属网，这大大增强了空间的层次深度和动感。为了吸引顾客到商店后方，设计师专门在那里设置了首饰展示架和一个充满活力的红色弧形收款台。镜子墙与试衣间相互平行，实现了现代和传统之间的连接和平衡，并强化了 Ora 品牌独有的奢华风格。

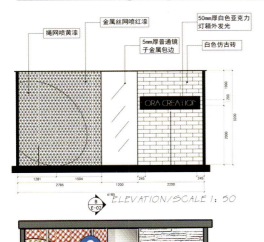

Elevation 4　立面图 4

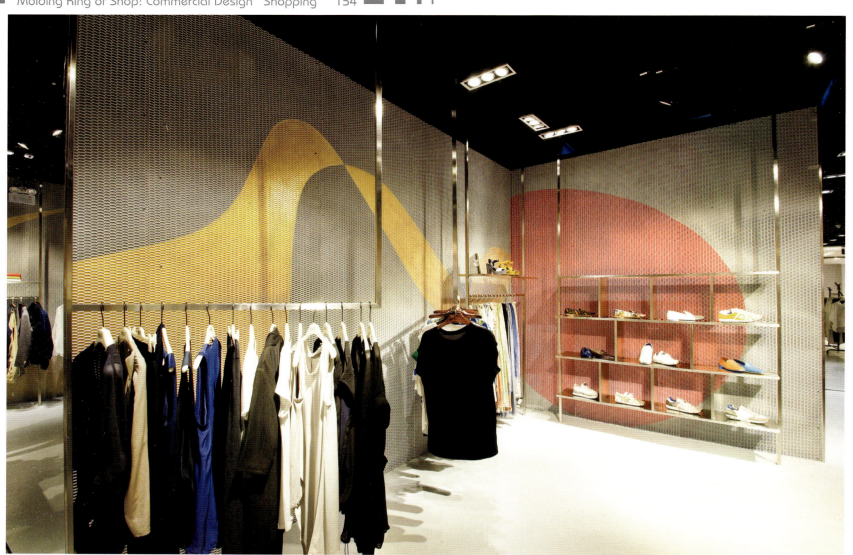

Karl Lagerfeld Store, Paris
卡尔·拉格菲尔德巴黎概念店

Design agency : plajer & franz studio gbr
Location : Paris, France
Area : 209 m²

设计单位：plajer & franz 设计工作室
项目地点：法国巴黎
项目面积：209 平方米

The new retail concept has been developed under the designer Karl Lagerfeld created and implemented by plajer & franz studio in cooperation with New York based creative agency Laird + Partners. Spread on 209 sqm shopping space, the overall store concept is inspired by the vision of the designer.
The main focus lies on reduction, black and white as well as strong contrasts in shapes and surfaces, blending edgy modern and classic elements. Materials and their surfaces play a vital role in the shop design, often creating multidimensionality. Reflecting surfaces allow playing with perspectives and thus the creation of subtle visual illusions. The lighting concept is inter-coordinated with the overall store design and integrated into the product display system. Besides its function to highlight the products, light is one of the principal design elements in store.
The store concept is centered on an exciting and elaborate digital experience – the virtual window to Karl Lagerfeld. Built-in touch-screens are placed in the fitting rooms encouraging consumers to capture their looks and share the pictures with friend via twitter, face book or email.

新概念店由设计师卡尔·拉格菲尔德亲自指导，由设计团队 plajer & franz 设计工作室和纽约设计事务所 Laird + Partners 联合打造。面积达 209 平方米，店内标志性的元素灵感来自于设计师卡尔·拉格菲尔德本人。

本案设计以简约为主，黑色与白色搭配，各种形状和表面形成鲜明的对比，前卫现代元素与古典元素融合。材料和表面在店铺设计中起着重要的作用，创造出一个多维度的店面，反射表面营造出微妙的视错觉。照明设计与整体商店设计一致，并且融入产品展示系统中，除了突出商品，照明也是该店设计的主要元素之一。

店铺的设计围绕着令人激动且精心打造的数码体验——从虚拟窗口走进卡尔·拉格菲尔德概念店内展开。通过在试衣间的内置触摸屏，顾客有机会捕捉它们不同样子的照片，并通过脸书、推特和电子邮件与朋友分享。

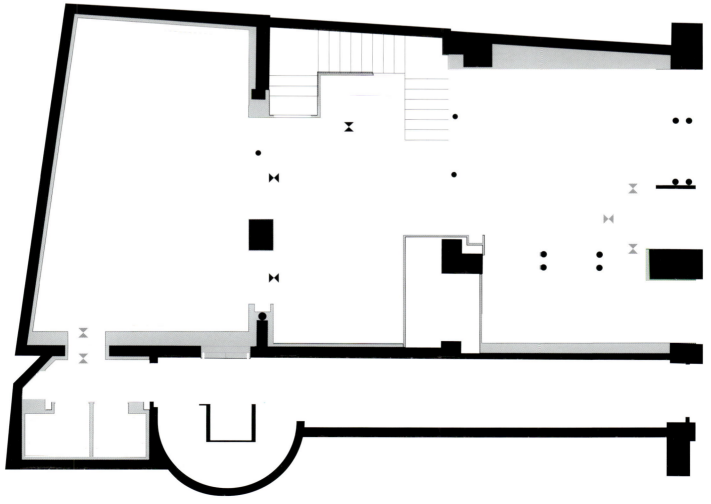

1F Plan　一层平面图

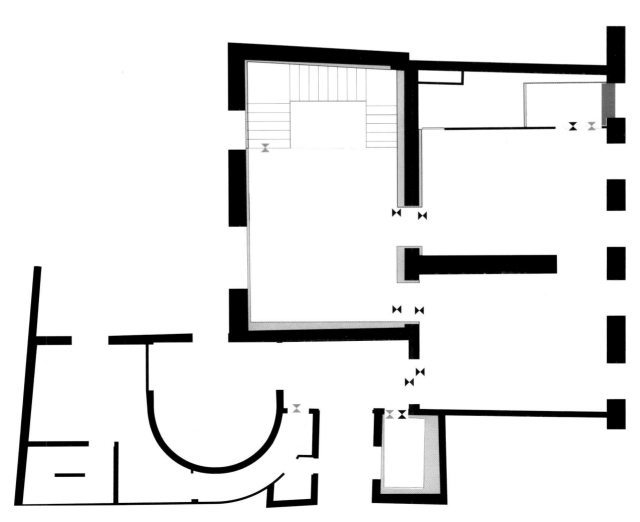

2F Plan 二层平面图

Hemu Brand Store
荷木服装品牌专卖店

Design agency : Yiduan Shanghai Interior Design
Designer : Xu Xujun
Location : Shanghai, China
Area : 60 m²

设计单位：上海亿端室内设计有限公司
设计师：徐旭俊
项目地点：中国上海
项目面积：60 平方米

Hemu, a high-end clothing brand with oriental style, sticks to the culture and aesthetics and tries to release the real personality with a concise and decent attitude.

Designer adopts simple and pure colors to biuld up a clean and elegant space, emphasizing the naive and simple mood. The grey color of bricks in large areas and coffee color of old trees are the main colors in this space, decorated with cement's different shades of grey. This fresh and simple style reminds of Zen, while this is enhanced by the steel bars with cool quality, making people have a strong impression of peace and quiet. Thus dark colored tone is finely integrated with the soft light clothing made by cotton and linen.

Decorations like dry lotus branch, vase, wood pile and books are functioning as key elements to deepen the sense of Zen. The design does not only aim at capturing the emotional request of consumers, but also is a meaningful practice and exploration into the creation of combining local culture with modern stores.

Elevation 立面图

"荷木"是一个集聚东方元素的高档服装品牌,坚持文化性、美学性、低调、大繁至简的设计态度,从心出发,回归心灵,溯源求本,呈现真我。

设计师用材朴素简单,色彩素墨淡雅,诉说着一种简单、淡泊的心境。以大面积青砖的深灰和老木的咖啡色为基调,再以水泥不同深浅的灰调为辅调,这样清爽、简约的色调让禅风呼之欲出。再融以线条简约的钢条,以冷、酷的质地进一步强化了"禅"所讲究的冷静、清心。这种硬质的深色基调与以棉、麻为原料的浅色系的软性材质的服装产生了很好的自然融合。

利用干荷枝、陶罐、老木桩、书籍等为装饰,起到了画龙点睛的作用,使"禅"文化得到进一步的延伸。荷木服装店的设计,不仅是对消费者内心情感诉求的捕捉,更是对现代购物场所如何融入地域文化的设计创新进行了一场有益的实践与探索。

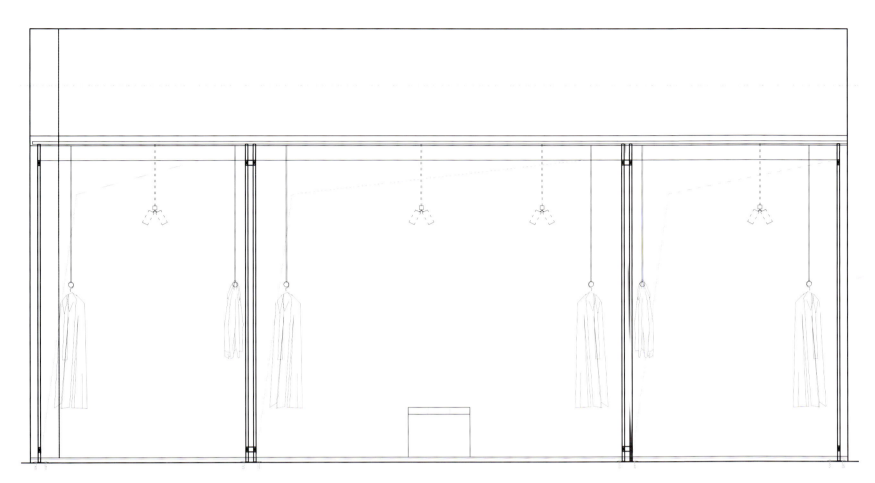

Detail 细节图

Ocean For in Huizhou

惠州四海名店

Design agency : Chu Chih-Kang Space Design
Designer : Chu Chih-Kang
Location : Huizhou, Guangdong, China
Area : 290 m²

设计单位：朱志康空间规划
设计师：朱志康
项目地点：中国广东惠州
项目面积：290 平方米

The client wants the stores to be wide as the Ocean and can spread around China like waves, so the design takes ocean as the theme to make sparking surface of sea by using specific materials and lights.
Stone and copper are used in the ground floor to create a sense of gorgeous and luxury, while the products displaying area is scattered with the reception area to break the traditional layout, and thus adding some kind of elegance in such a maze-like space. The wall on upper floor simulates the golden waves when sunshine irradiates on the surface of the ocean, while lighting variations are created by the reflection of light and mirror. Geometric display table looks just like the shining rock on the surface of ocean, turning into the focus in this area and echoing with the products as well as building up a luxury fashion atmosphere.

业主期望品牌版图如大海般辽阔,浪潮席卷全国,设计就此发想,以海洋为意象,运用材质及灯光,于空间中创造出波光粼粼的海面风光。

一楼以石材、黄铜营造出高级精品店的雍容气度,商品陈列区与顾客接待区交置错落,打破传统精品店对称的格局配置,在优雅中增添迷宫般的空间趣味。二楼墙面模拟阳光洒落海面的金色波纹,灯光与镜面折射,创造出绚丽的光影变化。几何切割的展台,仿若海面上闪耀的礁石,转化为空间的视觉焦点,与品牌单品相映生辉,渲染出奢华的时尚氛围。

塑造商铺之王：商业店面设计 购物篇 169

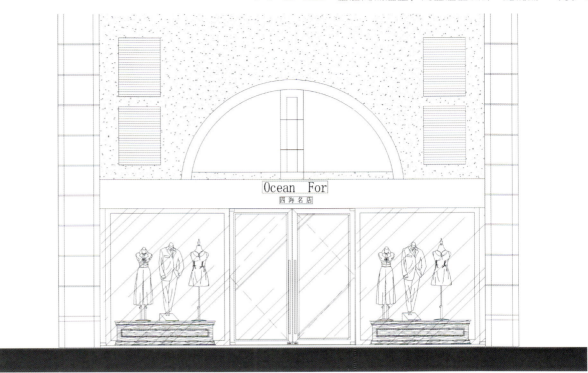

Elevation 立面图

1F Plan 一层平面图

1. 收银台　　2. 更衣室　　3. 储藏室　　4. 洗手间

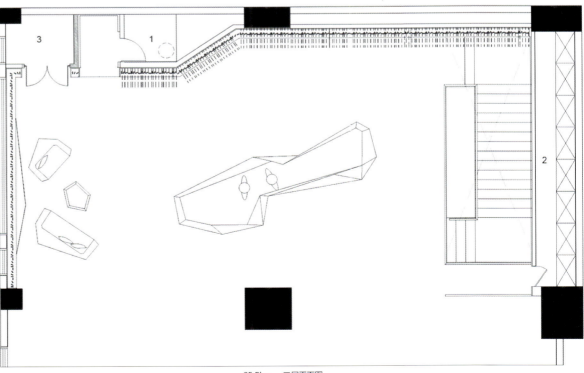

2F Plan 二层平面图

1. 更衣室　　2. 储藏室　　3. 佛堂

品质鞋包行天下
Luggage and shoes comforting your travel

鞋子、箱包最是体现品位的物件。细节处窥视端倪，极具品位之人必定不容丝毫瑕疵，行走人生的旅途，当然不局限于这一小小的空间之内。

Shoes and luggage are the ones expressing people's taste, which can never be ignored by people who have good tastes, since details show tastes completely. What you need during your travel of life is far beyond these small spaces.

VINTAGE STAR

Cat Bag
猫猫包潮流店

Design agency : AM Asociados
Designers : Marcelino Arranz, Maisa Mesa
Location : Avenida Gaudí, Barcelona
Area : 93.5 m²

设计单位：AM 建筑设计事务所
设计师：马塞利诺·阿兰茨、曼沙·梅萨
项目地点：巴塞罗那高迪大道
项目面积：93.5 平方米

The establishment highlights by showing an industrial, urban, airy and spacious concept, with exhibitions of different brands through corners. The designers recovered antique store spaces, adapting parts of the roof that are integrated, as owners wanted. The designers paved the room with laminate parquet which matches with the natural state walls, infusing an industrial environment throughout the premises.
Each corner has differentiated exhibitors following the design of each brand. Thus, the furniture in one of the corners has a white lacquered cabinet with shelves and drawers, in another one the designers have implanted metal furniture with spikes, for hanging products, and the third is made up of several wooden shelves in similar tone to the floor pavement.The window display consists of a wooden platform and guides embedded in walls to place the products and signage implementation with brand image.

SECTION AA'

SECTION BB'

SECTION CC'   **SECTION DD'**

PLANTA Escala 1:100

本案旨在建立一个工业化风格的、宽敞通风的市区店面，以在室内的各个角落展示各种品牌产品。设计师按照业主的意愿，修复了复古的店内空间，来呼应屋顶部分。店铺的地面采用了与墙面相匹配的木板，为空间注入工业的气息。

每个角落的立面设计都与其陈设的产品相对应。其中，一侧的陈设是白色漆柜，带有架子和抽屉；另一侧则是有长钉的金属架子，用来悬挂商品；还有个角落的陈设采用了与地板同色调的木架子和木地板。橱窗展示部分由一个木板平台和指示牌构成，嵌入墙体，不仅可以摆放产品，还能用于展示品牌形象。

Apple & Pie Children-shoe Boutique
apple & pie 童鞋专卖店

Design agency : Stefano Tordiglione Design Ltd
Designer : Stefano Tordiglione
Location : Hong Kong, China
Area : 85 m²
Photography : Stefano Tordiglione Design Ltd

设计单位：Stefano Tordiglione 设计事务所
设计师：Stefano Tordiglione
项目地点：中国香港
项目面积：85 平方米
摄影：Stefano Tordiglione 设计事务所

Walking into children's shoe store apple & pie is like entering another world – the giant apple that crowns the doorway hinting at the many delights that lie within. Inspired by the ethos behind the brand's name and its half apple-half pie logo, Stefano Tordiglione Design's concept combines the wellbeing elements represented by the apple with the more playful pie. The former is reflected in the use of environmentally- and child-friendly materials with a focus on wood as opposed to plastic for the furnishings, while the latter can be seen in the whimsical interior design which ranges from bright red apple-shaped sofas to imaginative wall displays. There is also practicality behind each well-thought out element. The seating hides storage space, while a lively tree design on one wall with white and red apples hanging from its branches, and elsewhere pie-shaped lattices and mounted fruit palettes, offer ideal shelving opportunities. In the windows, semi-circular pie-like features bring the logo and brand name full circle while providing window-shoppers with a taste of the various European shoe brands that can be found within.

For children coming to try and buy shoes, the back of the store offers a table at which they can sit and play between fittings, not far from a wall that provides familiarity through its giant blackboard design. Yet the focus is not solely on a positive experience for children. The iconic Kartell chairs surrounding the low table, the Ethel lighting hanging from the ceiling above, and the Giant Red Lamp designed by Anglepoise are design features which lend a sophisticated and elegant air to the store. The store effectively and effortlessly moves between the distinct worlds of parents and children , coupled with a combination of smooth curves and clean lines, above a warm wood-lined floor and with a color palette that blends bold reds and vivid greens with a calming mint and light beige.

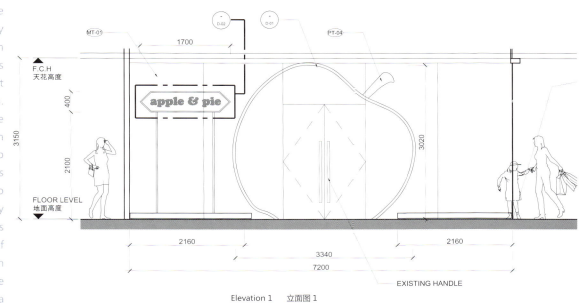

Elevation 1　立面图 1

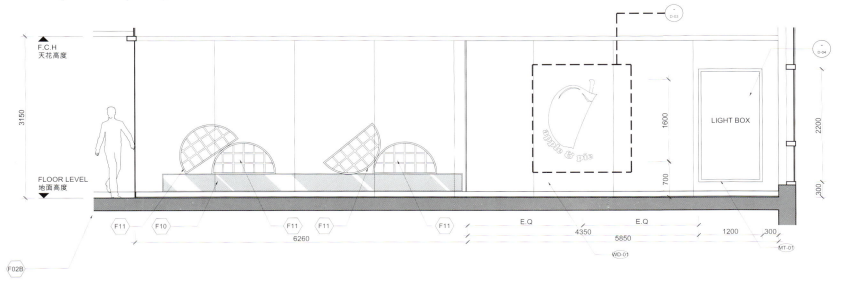

Elevation 2　立面图 2

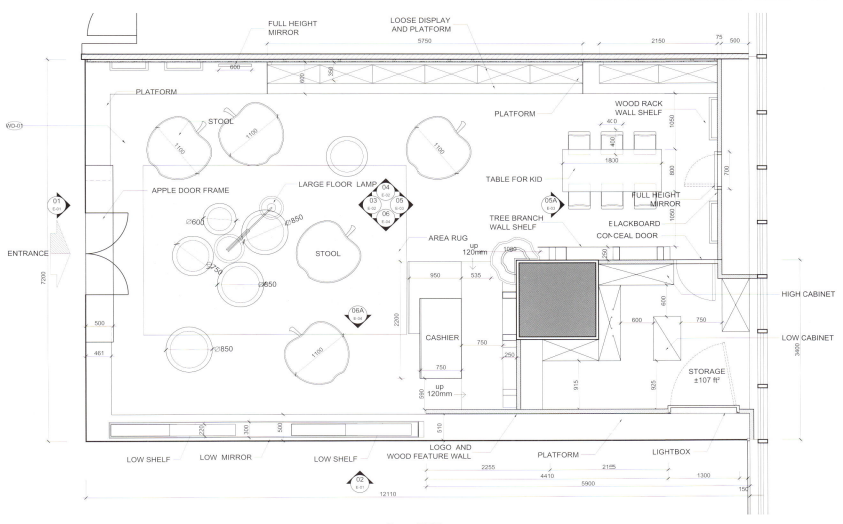

Plan 平面图

走进 apple & pie 童鞋专卖店就像走进了另外一个世界——贯穿门口的巨型红苹果隐约地暗示了店内的许多趣味。受到品牌名字以及其半苹果半馅饼商标的启发，Stefano Tordiglione 设计事务所的空间设计理念融合了苹果代表的健康元素以及馅饼的趣味性。苹果的健康含义从使用环保物料和有益儿童的物料中就能体现出来。店中运用了大量的木质材料，并尽量减少塑料材料的使用。馅饼的趣味在这个异想天开的空间设计中展示出来，包括了苹果形状的大红沙发和充满想象力的墙身展示。另外，每个设计元素都充分发挥其实用功效。红苹果沙发内含收纳空间。生动的苹果树上长满红色和白色的果实，可爱的馅饼造型货架放在孩子伸手可及的地方，这些元素都充分发挥着展示产品的功能。橱窗里，馅饼造型货架让品牌的名字和商标结合在一起，游人在店外就能观赏到各个欧洲品牌的鞋子。

对于年轻的小客人，店的后面设计了一面巨型的黑板，让孩子觉得熟悉舒服。黑板旁边放了一张小桌子，小孩可以在试鞋的同时休息和玩游戏。但是，店的设计并不只面向孩子。小桌子旁边的 Kartell 椅子，天花上垂下来的 Ethel 吊灯，以及 Anglepoise 设计的巨型红地灯，为店铺增添了典雅高贵的气质。整个店由柔顺的曲线和简洁的直线组合而成，地板也采用柔和色调的木板，再加上相似和对比颜色的交替使用，例如薄荷绿和米白色的柔和，相对于艳红和翠绿的对比，都让店铺轻松地游走于小孩和大人的世界中。

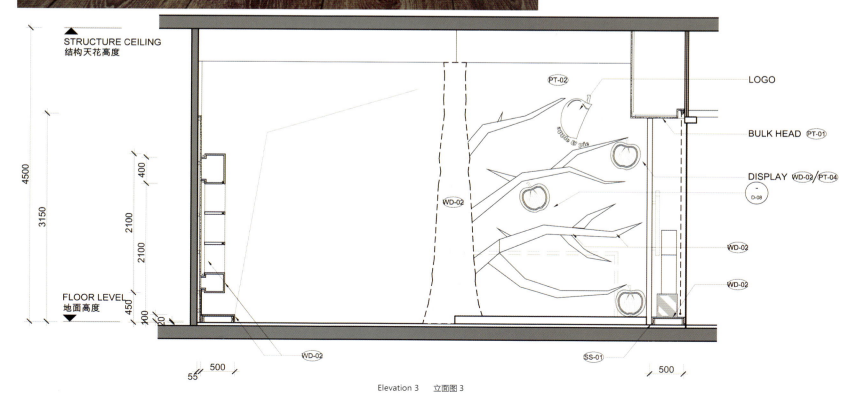

Elevation 3　立面图 3

Les Malles Moynat

Les Malles Moynat 箱包店

Design agency: Curiosity
Location: Paris, France

设计单位：Curiosity 设计事务所
项目地点：法国巴黎

The heritage of the original brand has been given new meaning, transforming the curved forms and unique details into contemporary leather goods. The challenge is to create a bridge between the brand's rich heritage and the modern world of luxury. This was a unique opportunity to create a coherent message in the architectural design of the new boutique as well as the brand's graphics, packaging and a limited edition book that presents the rich past of Moynat.

The first impression is a feeling of harmony, where one main element captures your attention. As you move through the space, it evokes curiosity to discover the history of the house and the excellence of the craftsmanship. Soft light emanates from a glass ceiling, reminiscent of the art deco period that influenced the style of Moynat. The floor, composed of a titanium-coated glass mosaic, reflects light with a water-like effect. While the ground floor is an open space, the mezzanine offers a more intimate environment, with a salon, a gallery and the men's collection.

原品牌遗留下来的东西已经被赋予了新意，将弯曲的外形和独特的细节运用在当代皮革制品上。设计中最大的挑战就是在品牌丰富的文化遗产与现代奢华的世界之间寻找到连接点。这是一个难得的机会，在新精品店的建筑设计，品牌形象与包装和讲述 Moynat 悠久历史的限量书籍间创建连续传达品牌的信息。

该店给人的第一印象就是和谐，只需要一个主元素就能够吸引过客的目光。漫步于整个空间，它将唤起你的好奇心去探索其间的历史痕迹和卓越的工艺。柔和的光线从玻璃天花板倾泻下来，让人想起那个深深影响 Moynat 的装饰艺术时代。地板采用钛镀膜镜面马赛克，像水面一样反射着光线。一楼是一个开放的空间，而夹层则是一个更为私密的环境，设有沙龙、走廊和男装区。

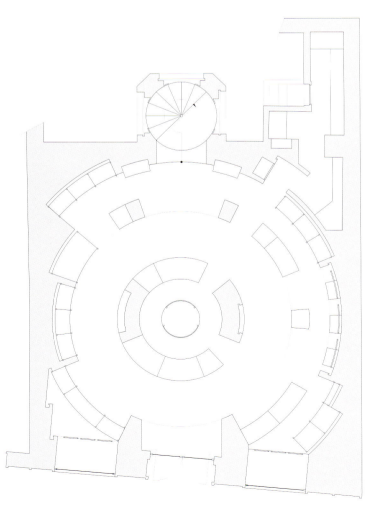

1F Plan　一楼平面图　　　　2F Plan　二楼平面图

Vintage Star – Jessica Simpson Collection

复古之星——杰西卡·辛普森品牌店

Design agency: Sergio Mannino Studio
Designers: Sergio Mannino, Francesca Scalettaris
Location: New York, USA

设计单位:塞尔吉·奥曼尼诺设计事务所
设计师:塞尔吉·奥曼尼诺、弗兰西斯卡·斯卡莱塔里斯
项目地点:美国纽约

When the owner initially approached Sergio Mannino Studios, the idea was to create a sophisticated boutique that would offer women incredible product at a great price. He wanted a clean, modern space with an intimate feel and "just a touch of vintage". Mr. Mannino accepted the challenge and just completed the first prototype store in Shanghai, adopting innovative modern elements mixed with old world glamour to create an aesthetic not often used in the world of retail design. Through the majestic windows, product appears framed in sharp geometric displays; grand yet warm, incorporating a freestanding chrome ottoman bench and other domestic objects to give the store a sophisticated and comfortable feel.

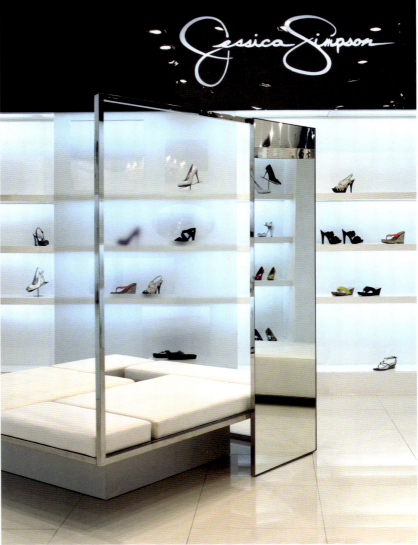

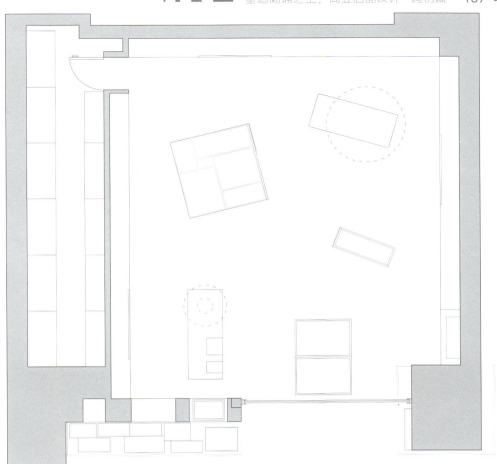

客户要求设计师打造一个精致的为女性顾客提供物美价廉商品的空间。这个空间必须同时符合干净、前卫但又不让人感觉拘谨的要求，同时也带有一点复古的味道。设计师接受了这个挑战，并刚在上海完成了第一家示范店，将创新性的现代元素融入复古的基调来营造在零售商店并不常见的独特的审美品位。通过华丽的橱窗，商品以醒目的几何框架展示出来，大气却不失温和。设计师运用独立的铬制绒垫长凳以及家庭用品，使商店更显精致、舒适。

Munich Flagship Store Barcelona

巴塞罗那慕尼驰旗舰店

Design agency : Dear design	设计单位：Dear 设计事务所
Designers : Ignasi Llauradó, Eric Dufourd	设计师：Ignasi Llauradó、Eric Dufourd
Location : Barcelona, Spain	项目地点：西班牙巴塞罗那
Area : 55 m²	项目面积：55 平方米
Photography : Lafotográfica	摄影：Lafotográfica

Dear design's concept for Munich flagship store, located within the L'illa Diagonal Shopping Mall, focuses on carefree movement and angular pieces. The challenge of this project was to convert this small space into an endless, extended and warm field. This premise was linked to the idea of emphasizing within the store the brand's character: freedom and uniqueness. The finishes of walls and ceiling create games of reflection to multiply and expand the space, the trees, the hills and the Munich logo. The irregularity of the space was used conveniently to make it become part of the concept, member of the game.

The surfaces are finished with black glass, mirrors, metallic trees and hanging jails for special trainers, enforcing the space's experimental, carefree and limitless atmosphere. The angular and clunky space with its hard edges and seemingly moving parts is clearly an attempt to say that the septuagenarian brand is nowhere near slowing down.

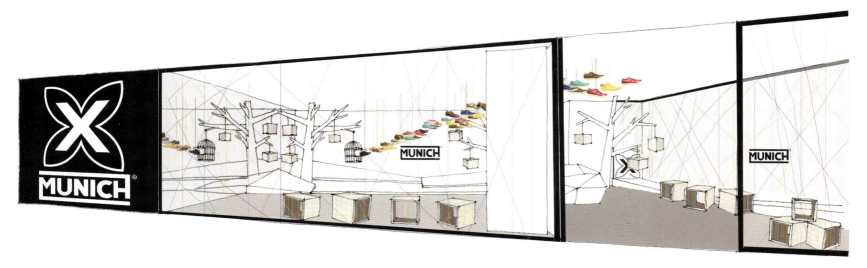

Elevation 1　立面图 1

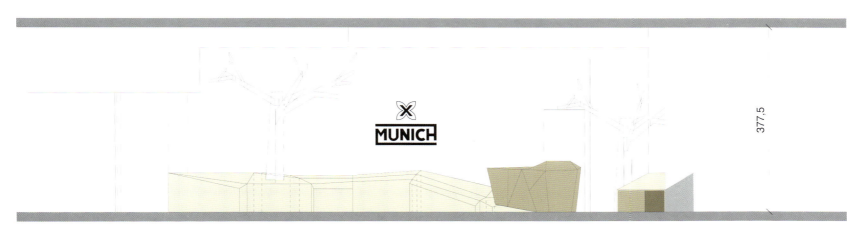

Elevation 2　立面图 2

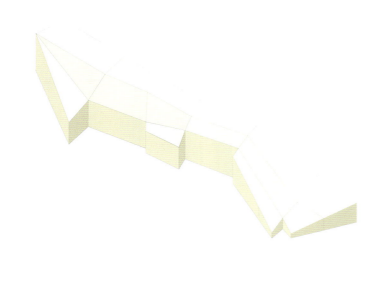

Detail 细节图

慕尼驰旗舰店由 Dear 设计事务所设计，坐落于利亚对角线购物中心，专营时尚运动鞋品。在该项目设计中，最大的难题就是将窄小的空间设计成为宽阔的、可延展的温馨空间，还需要与品牌的理念融为一体，诠释品牌自由和独特的个性。通过墙上的装饰和天花板的设计，创造出假山一般的展示台、树木的形象，配上慕尼驰的标识，整个空间获得了视觉上的扩展。空间的不规则性也是设计的一部分，增加店内的趣味。

店内陈设的表面由黑色玻璃镜面构成，金属树形结构和悬挂的鞋盒都增强了空间的体验性，营造出一种无拘无束的氛围。棱角分明的空间设计看似笨重，但实际上是在表明这个经历了七十年风雨的品牌并没有放慢追寻品质的脚步。

Vertical View 俯视图

熠熠珠宝盼生辉
Sparkling jewels shining your life

无论是璀璨的钻石，还是温润的玉石，抑或是闪亮的金银，都十分吸引人，让人为之倾心，爱不释手，让笑容也为之灿烂。

Dazzling diamond, humid jade, resplendent gold and silver are all attractive, making people fall in love with them without letting them go, and lighten their smiles.

Wellendorff Jewellery Boutique
华洛芙高级珠宝店

Design agency : Stefano Tordiglione Design Ltd
Designer : Stefano Tordiglione
Location : IFC Mall, Hong Kong, China
Area : 47 m²
Photography : Allan Leung

设计单位 : Stefano Tordiglione 设计事务所
设计师 : Stefano Tordiglione
项目地点 : 中国香港国际金融中心
项目面积 : 47 平方米
摄影 : Allan Leung

In a refined and elegant setting in the heart of IFC lies the new Wellendorff boutique, alongside the outstanding quality and beauty of the brand's jewellery items, inspired Stefano Tordiglione Design's concept.

Entering the soft-lit store, visitors are welcomed by warm wooden floors, their motif taken from the floors of old German castles where different cuts of wood were used to create patterns. The walls are pale grey with light; the columns are gracefully interspersed, lending a classical air to the decor. A subtle pattern appears across the walls, a magnified yet softened version of Wellendorff's floral identification. It goes across the ceiling and continues to the dark wood façade on the boutique's exterior.

Not immediately identifiable, as visitors go deeper into the store it forms an embrace, part of the concept of envelopment that continues throughout, further enhanced by the prevalence of curves and circles. The rounded display cases, the wave in the walls and the soft, circular cashier's desk etc, all of the design elements bring a sense of continuity to the concept. High quality materials are used across the store, reflective of the fine nature and attention to detail of the brand's products. The cashier's desk is partly covered in leather, while a large fireplace has been designed from two types of exquisite marble. The color palette is predominantly neutral.

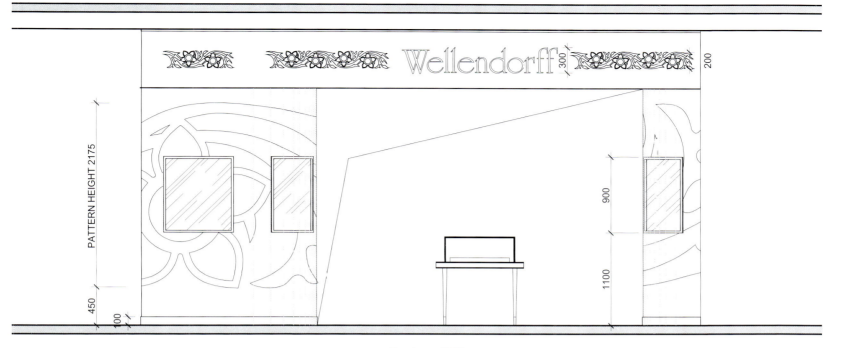

Elevation 立面图

全新华洛芙高级珠宝店位于香港国际金融中心的中央,高雅而脱俗。华洛芙珠宝质量卓越、美丽非凡,激发了 Stefano Tordiglione 设计事务所的设计灵感。

走进商店,顾客首先看到的是柔和的木地板。这个创作意念源自古老德国城堡的地板,木料被锯成各种形状,拼出不同的图案。浅灰色的墙壁配以灯光,加上优雅的柱子点缀,整个装饰充满古典气息。墙壁中央设有一个巧妙的图案,那是放大的华洛芙花卉标志,设计柔美,穿过天花板,一直延伸到珠宝店外面的黑木立面。

椭圆形的陈列柜、墙上的波纹、色调柔和的半圆形收银台等,一切设计元素给人一种一气呵成的感觉。整个店铺采用优质材料进行装饰,反映出品牌产品的优良特质和细节。收银台部分由皮革包裹,而大壁炉由两种精美的大理石打造,以中性色泽为主调。

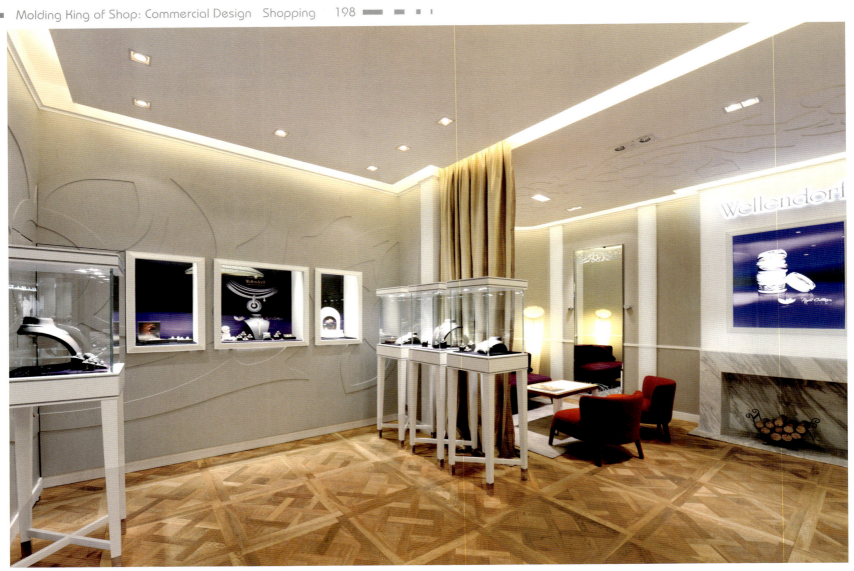

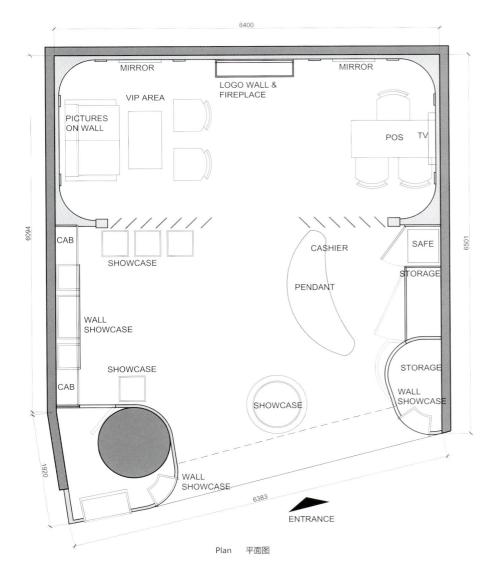

Plan 平面图

Hirsh Jewelry Boutique, London

伦敦赫什珠宝精品店

Design agency : Diego Bortolato Architetto
Designer : Diego Bortolato
Location : London, UK
Area : 60 m²
Photography : Daniela Bortolato

设计单位：迪亚戈·博尔托拉托建筑事务所
设计师：迪亚戈·博尔托拉托
项目地点：英国伦敦
项目面积：60 平方米
摄影：丹妮拉·博尔托拉托

The new Hirsh Jewelry Boutique that the designers have developed in London, Grafton St. 13, is for the Hirsh Family, Specialists in colored diamonds and unusual gemstones. In the middle of the jewels empire of the Bond St. Area, the furniture of the store is designed to define a new character specific to the brand Hirsh. The furnishings are all designed specifically for this store, developed and manufactured in Italy, characterized by the use of materials and technologies of the scope of museology.

Elevation 1　　立面图 1

位于伦敦市格拉夫顿街，迪亚戈为赫什珠宝设计了新精品店，该店专营彩色钻石和稀有宝石。在恍如珠宝王国的邦德街中间，该店的家具设计为赫什珠宝打造了一个全新的形象。所有的陈设都是在意大利专门定制生产的，通过材料的使用和做工技巧，来彰显品牌的魅力。

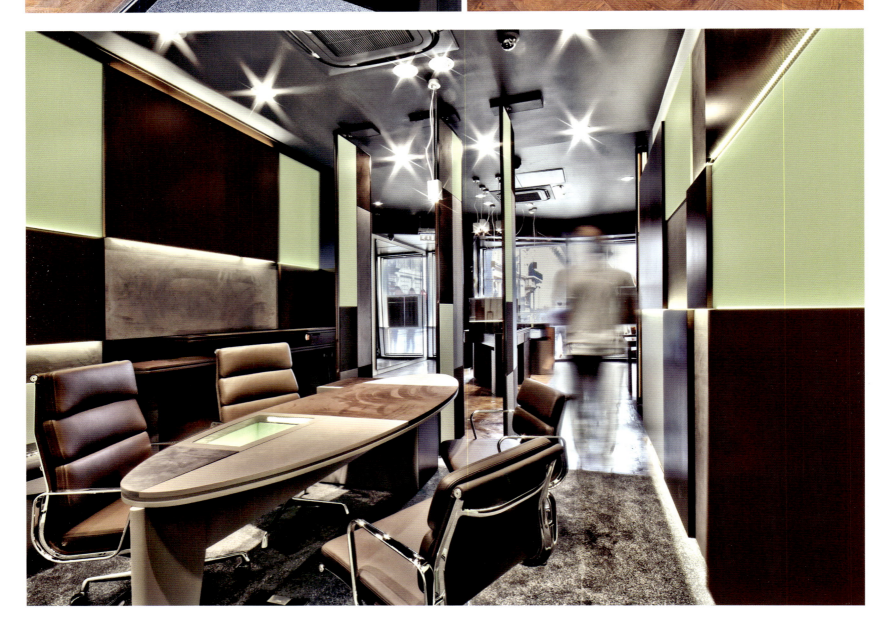

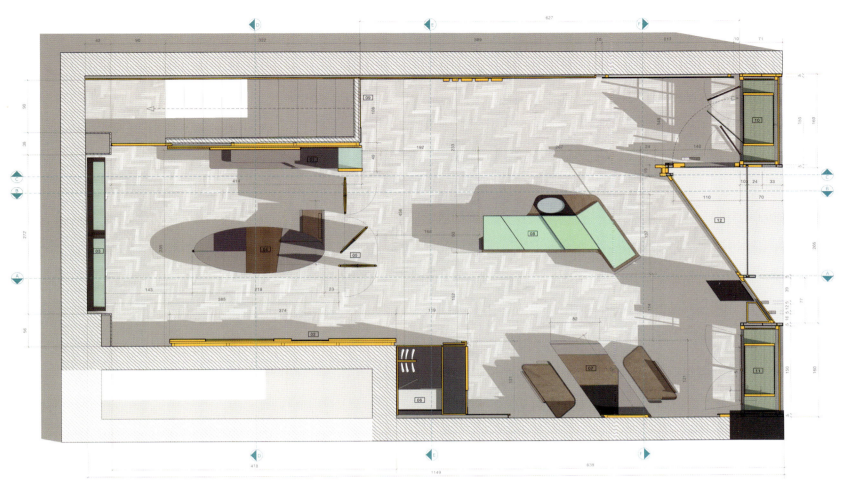

Plan 平面图

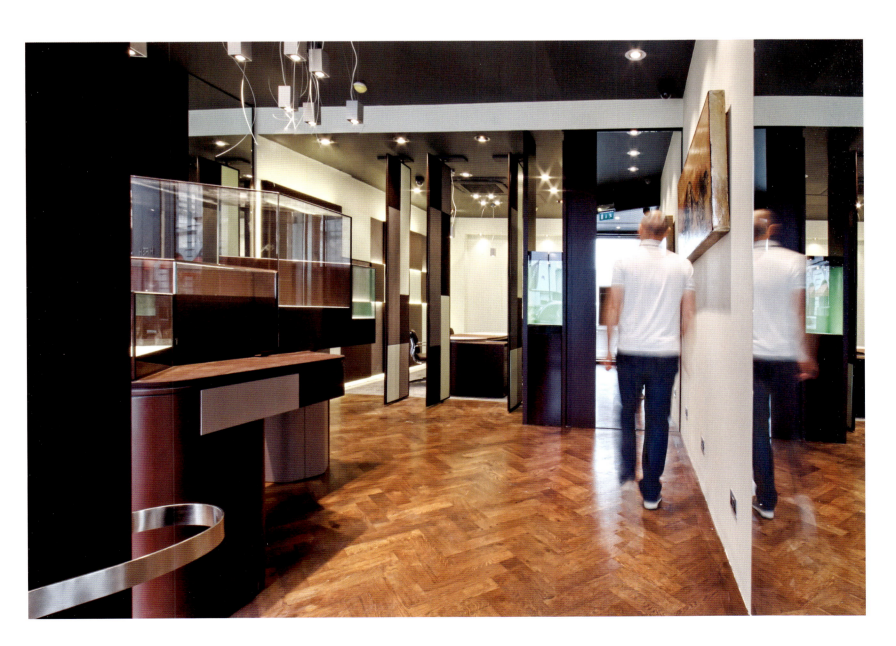

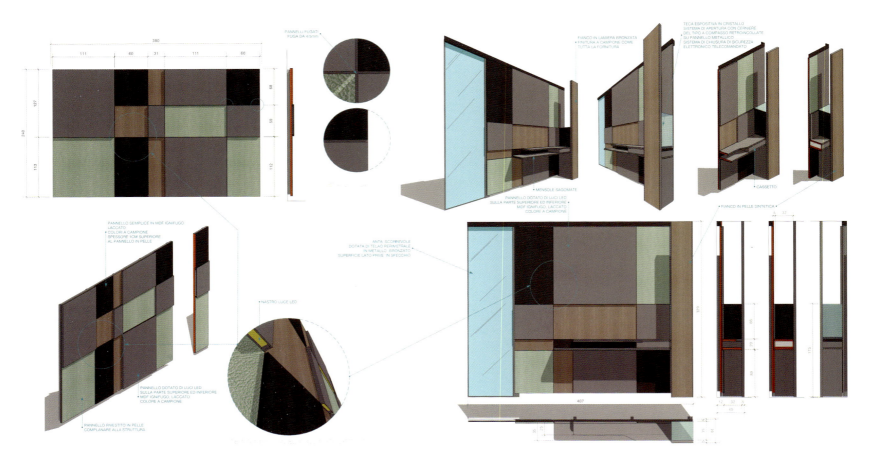

Elevation 2　立面图 2

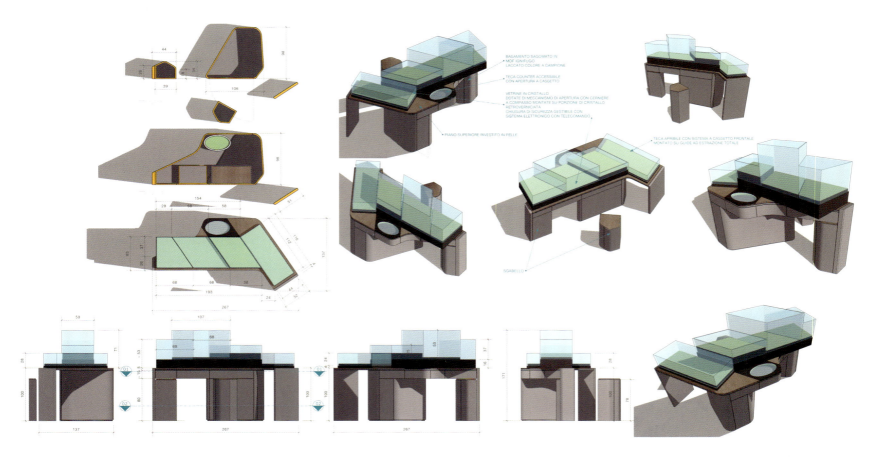

Detail 1　细节图 1

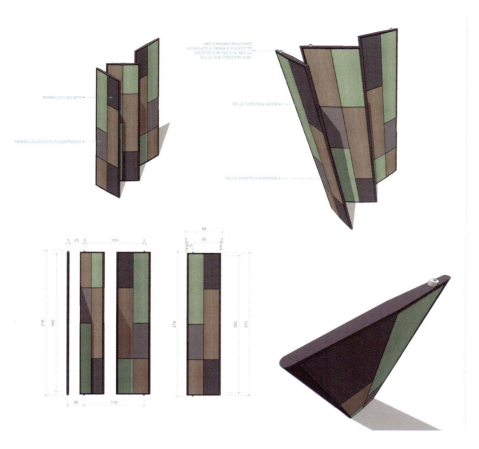
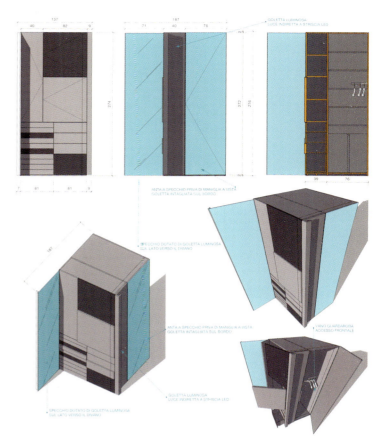

Detail 2　细节图 2

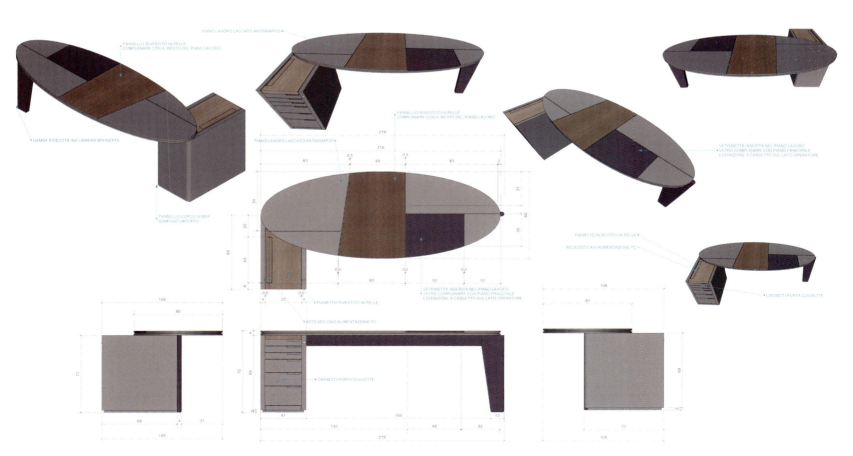

Detail 3　细节图 3

Dinh Van Jewellery Boutique

Dinh Van 珠宝精品店

Design agency : Stefano Tordiglione Design Ltd
Designer : Stefano Tordiglione
Location : Hong Kong, China
Area : 42 m²
Photography : Dinh Van Paris

设计单位：Stefano Tordiglione 设计事务所
设计师：Stefano Tordiglione
项目地点：中国香港
项目面积：42 平方米
摄影：Dinh Van Paris

Dinh Van is a brand renowned across in France, in particular in Paris, and Stefano Tordiglione Design sought to play on this fact by referencing landmarks of the French capital throughout the store. A charcoal grey and white pattern makes an impact on the feature wall, depicting part of an iron architectural structure. It is immediately reminiscent of the Eiffel Tower, transporting visitors to Paris. On other parts of the wall, black and white prints once used in a Dinh Van advertising campaign show off the city's Pantheon and Place de la Concorde, adding to the cocooning effect of the design concept.

The display cabinets are particularly outstanding, taking the Parisian notions one step further. Here the bases of these furniture pieces resemble the famous Tower itself, created from slender rods of metal, tapering towards the displays cases and standing as miniature cross-sections of the Tower. They are thoughtfully arranged in a store where space is at a premium. Yet even so, the design team has risen to the challenge and expertly incorporated an indented sales area in addition to a compact cashier table.

From the stylish interior – where the wall patterns highlight certain areas of the shop and point to particular jewellery displays – the theme continues outside the store and onto the metal clad exterior where the dark grey pattern is more subtle against the pale metallic facade, however, the result is just as effective as before. Graphic and contemporary it represents the brand and its values immediately manages to evoke a sense of glamour and elegance at the same time.

Dinh Van 品牌在法国各地很有名，尤其是在巴黎。于是，Stefano Tordiglione 设计事务所充分利用这个特点，使整个珠宝店的设计参照法国首都巴黎的标志性建筑物。炭灰色和白色图案是其墙壁的特色，描绘出铁艺建筑结构的一部分，这立即使人联想起巴黎的埃菲尔铁塔，吸引游客前往巴黎。墙身的其他部分是黑白相间的印刷品，它们都曾用于 Dinh Van 的广告活动，展示巴黎的万神殿和协和广场。

陈列柜更为醒目，进一步展示了巴黎的理念。这里，每件家具的基座都模仿著名铁塔，由细长金属条做成，向陈列柜方向逐渐变细，就像是铁塔横截面的微型图。这些家具都精心放置在珠宝店内，但设计师团队敢于挑战，除了设有紧凑的收银台外，还巧妙地设计了一个相互交错的销售区域。

店面室内装饰时尚，其墙壁图案突出商店的独特区域和特别的珠宝陈列，该主题一直延伸到店外和包有金属装饰的外立面。外围装饰了深灰色图案，在暗淡的金属门面的反衬下更显精细，虽然含蓄，但其效果与以前相若，因为它代表了品牌的形象和现代感，极具魅力、极为优雅，充分体现了品牌价值。

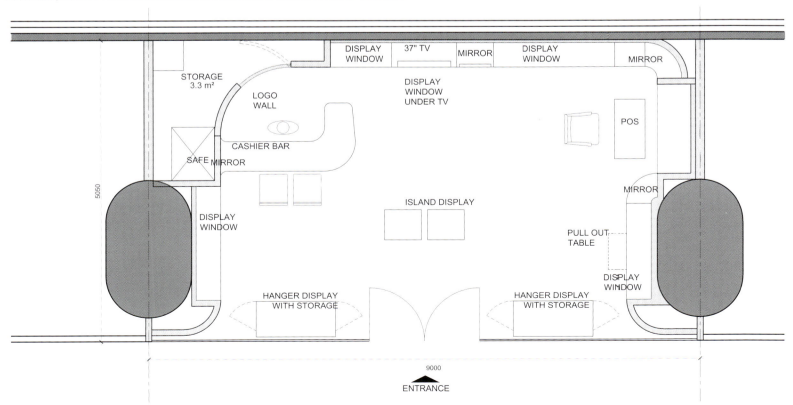

Plan　平面图

Leo Pizzo Boutique Milan
米兰利奥皮佐精品店

Design agency : Diego Bortolato Architetto, Gianluca Re Architect
Designer : Diego Bortolato
Location : Milan, Italy
Photography : Jurgen Eheim

设计单位：迪亚戈·博尔托拉托建筑事务所、詹卢卡建筑事务所
设计师：迪亚戈·博尔托拉托
项目地点：意大利米兰
摄影：尤尔根·伊罕

The project of the Leo Pizzo Boutique in Milan is a case the designer has developed with his colleague Gianluca Re Architect from his own town. This is the first monobrand shop that the client, Mr. Leo Pizzo, has opened on the fashion town, Milan. They used natural material, the same of the historical Villa Necchi: natural piedmont walnut (one single old tree for the whole furniture) back painted glass and bronze. The space is developed as a museum, quite clean, and with respect with the original structures.

米兰利奥皮佐精品店是由迪亚戈·博尔托拉托与其同乡同僚吉安路卡·雷一起设计的。这也是本案客户利奥·皮佐在米兰这个时尚之城开设的第一家品牌店，运用了米兰内基别墅所使用的天然材料：皮埃蒙特核桃木（所有家具由一棵树制造而成）、背漆玻璃和青铜。该空间就像一个博物馆，保持着最原始的结构，安静又整洁。

STATO DI FATTO　　　　　　　　　　　　　　SITUAZIONE DI PROGETTO

Elevation 1　立面图 1

Elevation 2　立面图 2

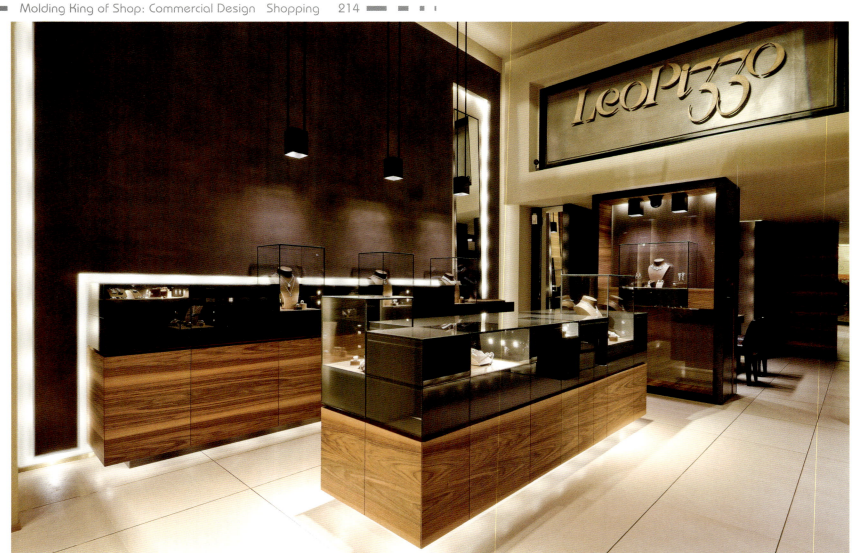

Valuable Flagship Store, Suzhou
万宝缘银楼苏州旗舰店

Design agency : Zhumu Decoration& design Co., Ltd.
Designer : Chen Jie
Location : Suzhou, Jiangsu, China

设计单位：上海筑木空间装饰有限公司
设计师：陈洁
项目地点：中国江苏苏州

The entire space adopts red to be the main tone decorated with golden, to create a cheering atmosphere with Chinese traditional style. Walls around the store are decorated with large oil paintings instead of light boxes and spray painting techniques, to make a visual impact and thus promote the quality in this area in large degree. Crystal lamps above make the space luxury and gorgeous and they are matched with the shining silverwares, making people appeal to those products displaying. The layout of the cabinets is concise and simple to make it convenient for consumers to pick up what they like. People always lose their control in such a luxury space with a hint of oriental style with adorable silverwares full of their eyes, isn't that a feast of eyes?

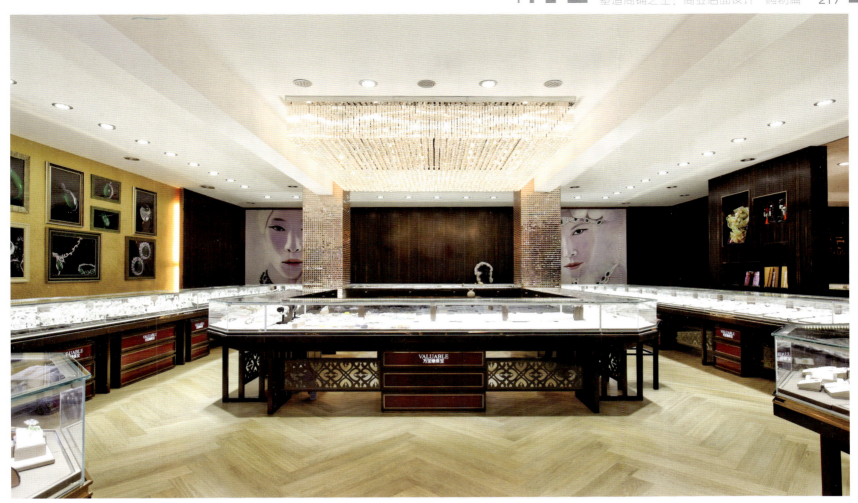

为营造中国传统的黄金珠宝店喜庆的店面效果，本案整体采用酒红色色调，并辅以玫瑰金。店内所有的墙面一改传统的灯箱片及喷绘的处理手法，而改用大幅油画装饰，具有极大的视觉冲击力，并在很大程度上，提升了装修的档次。水晶吊灯使得空间显得奢华高雅，搭配闪亮的银饰，让人对这些精细高雅的饰品爱不释手。柜台的格局简洁分明，为顾客挑选喜爱的饰品提供了方便。在这样一个带着些许中国风的奢华空间内，人们总是没有丝毫抵抗力，沉浸在这样一个雅致的空间里，满眼尽是赏心悦目的饰品，何尝不是一场视觉的盛宴呢？

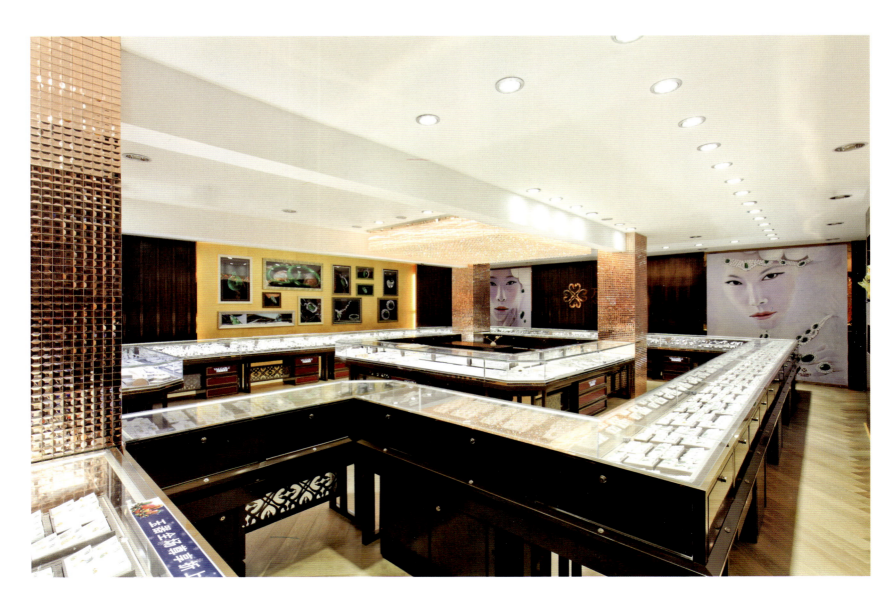

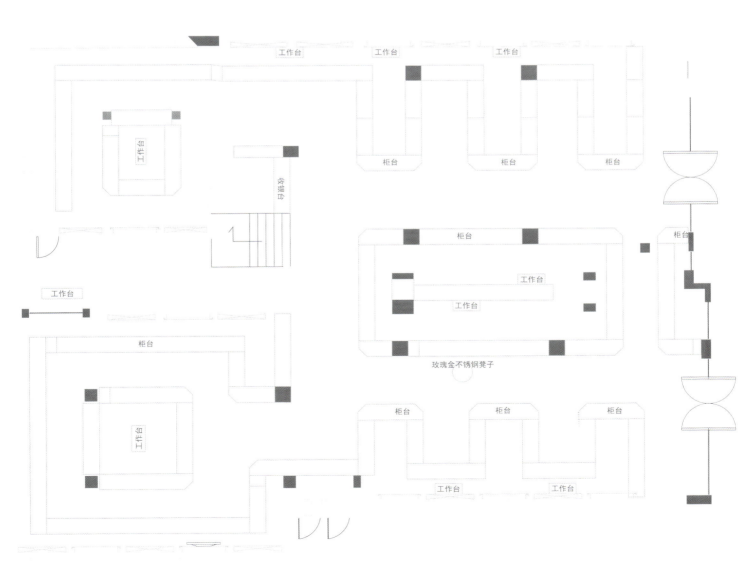

Plan 平面图

美容美发换新颜
Cosmetology and hairdressing renewing your life

换发型、做SPA，享受公主般的待遇，是每个爱美女士憧憬的理想生活，闲暇日子坐在这些别具一格的小店里，容光焕发地接受目光的洗礼。

A comfortable spa or hairdressing like a princess is the favorite of every lady, you could just sit in such an exquisite room and wait to be a charming fair lady appealing to all jealous sights.

Pretty Beauty Health Spa

汕头唯美伊人养生美容馆

Design agency : Shantou Youxin Decorative Engineering Co., Ltd.
Designers : Wu Miaochang, Lu Zhonglian, Wu Qiaopeng
Location : Shantou, Guangdong, China
Area : 265 m²
Photography : Wang Yuxun

设计单位：汕头市友信装饰工程有限公司
设计师：吴淼昌、卢中炼、吴乔鹏
项目地点：中国广东汕头
项目面积：265 平方米
摄影：王毓洵

This case takes red as the main tone, while Logo of this brand is interpreted finely with Chinese signet, emphasizing the new concept of health by combining the traditional elements like Chinese furnishing, classic patterns, caved flowers and classic lamps. The first floor is experience area with lounge, dressing room, spa room, physiotherapy room and bathroom, etc. The decorations here are not limited in one style, but being both in western and oriental style to fit this space, creating a balance between all rooms. Chinese charm, European attraction and the beauties of Southeast Asian style are put into this place at one time; this could avoid aesthetic fatigue caused by single united style. in such way, a quiet, ease and comfortable warm space is built to make the customer indulge in pleasures without stop.

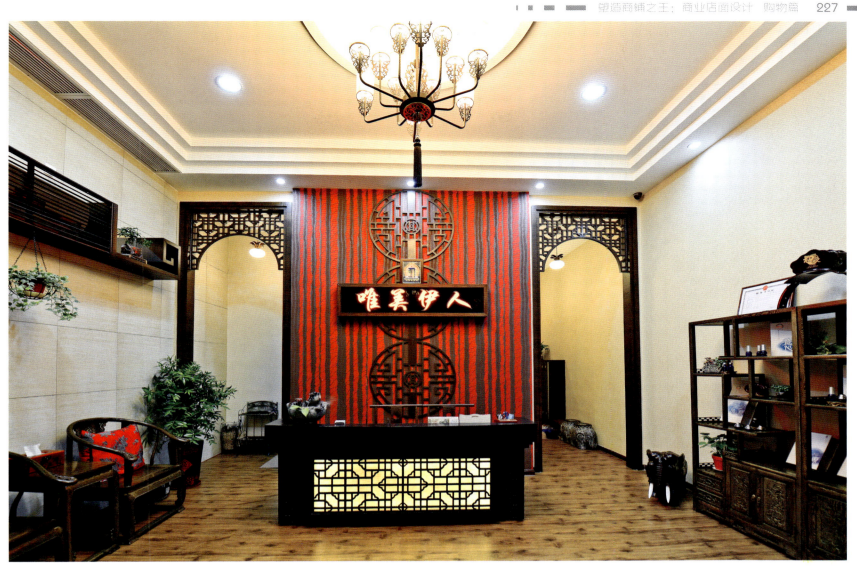

1F Plan　一楼平面图

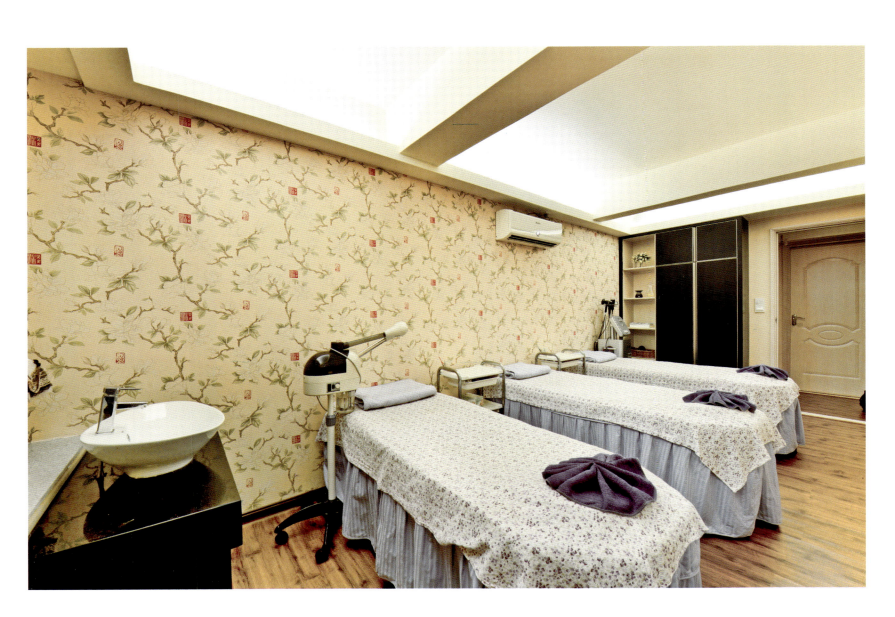

店面装饰以中国红为主色调，品牌LOGO巧妙结合"中国印章"的表达形式，通过中式家具、传统纹样、雕刻通花、古典灯具等东方元素，方圆融合来诠释养生主题新概念。二楼为体验区域，配有休息区、更衣室、SPA房、美容理疗房、淋浴间等，装饰形式不拘一格。通过运用混搭手法融汇东西方审美观，整合协调各式区间，中式韵味、欧式情调、东南亚风情共冶一炉，避免过分强调风格统一而造成审美疲劳，旨在营造静谧、放松、舒适、温馨的环境。

2F Plan 二楼平面图

IL SALONE
IL SALONE 美发店

Design agency : Egue y Seta
Designers : Daniel Pérez, Felipe Araujo
Location : Barcelona, Spain
Photography : Victor Hugo

设计单位：Egue y Seta
设计师：丹尼尔•佩雷兹、菲利普•阿罗约
项目地点：西班牙巴塞罗那
摄影：维克多•雨果

From the outside, we can see through its transparent facade the inner side of a premise that far from hiding, aims to showcase an honest, devoted and artifice-less working style. The font of logo is very simple and retro, expressing the natural, unrefined brand concept.

Once in, we are greeted by a warm looking natural oak pavement. This " wooden carpet ", highly resistant and low maintenance, climbs the walls, creating a waist-high baseboard that hosts indirect lighting strips and serves, at the same time, as mounting surface for the minimal shelving units that make up each hairdressing station, along with a square mirror that floats over the vertical plane and a sleek ergonomic chair in dark chocolate upholstery. Sitting on it , rather a "guest" than a customer, and behind him, more than a hairdresser a hostess who will make you feel like home.

Heads washing seats, that should not resemble any piece of household furniture, are located at the back of the room, flanked by a wall on which the client, carefully placing two black and white enlarged photos of Sophia Loren and Macello Mastroianni, wanted to elegantly nod to her cultural background.

透过纯净通透的玻璃外立面内部情景一目了然，这种做法旨在展示一个完全真实的工作现场。LOGO采用极简的复古字体，传达出真实自然、不加雕琢的品牌理念。

一进入室内，温暖自然的橡木地板便映入眼帘，这种"木质地毯"防缩性强，维修成本低，并由地面延伸到墙面齐腰高的位置作为护壁板，木质结构内部的内藏灯具还有利于间接照明，同时，凸出的面板作为每个美发座席的置物台，台上挂有长方形的镜子，犹如漂浮在墙面上一般，坐在深咖啡色的、线条柔和的符合人体工学的椅子上，顾客如同回到了家一样轻松自在。

洗头席位在空间的最里侧，用一面墙进行遮挡，墙面上索菲亚·罗兰和马塞罗·马斯楚安尼巨大的黑白照片，传达出意式的优雅。

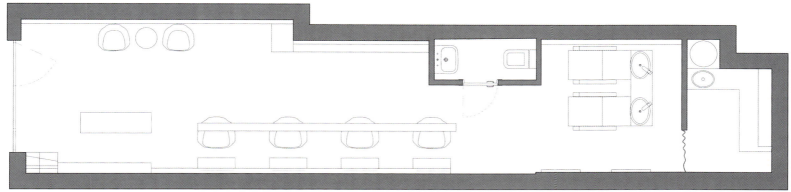
Vertical View　俯视图

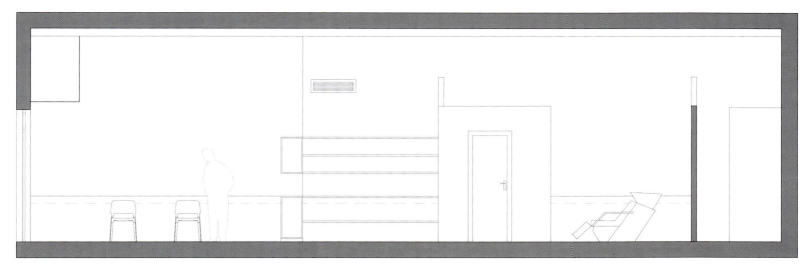
Elevation 1　立面图 1

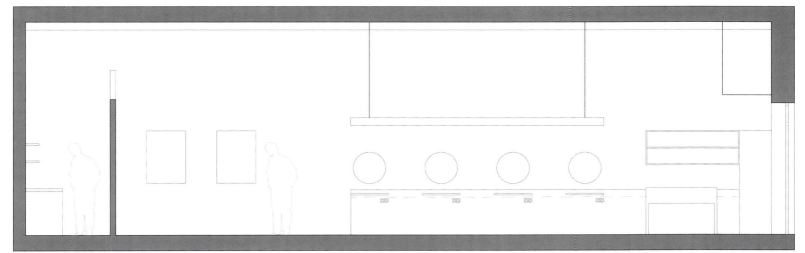
Elevation 2　立面图 2

Best Cuts and Colours

Best Cuts and Colours 美发店

Design agency : Studio Equator
Designers : Carlos Flores, Benjamin Fretard
Location : Melbourne, Australia
Area : 62 m²
Photography : Anne-Sophie Poirier

设计单位：Equator 设计事务所
设计师：卡洛斯·弗洛雷斯、本杰明·弗勒塔尔
项目地点：澳大利亚墨尔本
项目面积：62 平方米
摄影：安妮-索菲·普瓦里耶

Building from an existing brand, Best Cuts and Colours has opened another salon in the centre of one of the most popular shopping centers in Melbourne. This expanding brand has had over 10 years experience in the Hair & Beauty industry, so for their 15th store opening they found creative direction through Studio Equator.

The overall concept explains true simplicity and modernism with a stylish edge emphasizing fashion, beauty and quality. The façade of the salon opens with painted timber panels framing the signage and shelves decorated with white LED lights. A matte floor tile covers the space and contrasts off the glossy black and white interior, adding warmth to the fit out. The focal point in this design is the use of mirroring and lights that Studio Equator has played with in the ceiling installation. Light fittings then hang down at the same point next to each mirror panel and intern reflects through. The cutting/color stations are simple with ceiling to floor individual mirrors, leather black chairs and soft white painted walls.

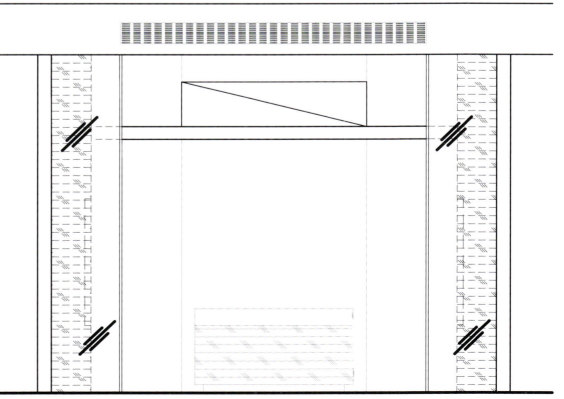

Elevation 1　立面图 1

The overall concept explains true simplicity and modernism with a stylish edge emphasizing fashion, beauty and quality. The façade of the salon opens with painted timber panels framing the signage and shelves decorated with white LED lights. A matte floor tile covers the space and contrasts off the glossy black and white interior, adding warmth to the fit out. The focal point in this design is the use of mirroring and lights that Studio Equator has played with in the ceiling installation. Light fittings then hang down at the same point next to each mirror panel and intern reflects through. The cutting/color stations are simple with ceiling to floor individual mirrors, leather black chairs and soft white painted walls.

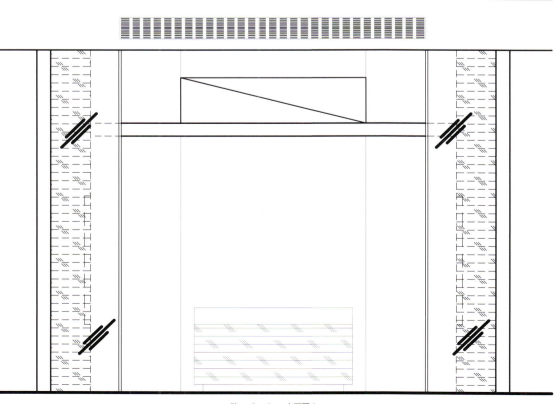

Elevation 1 立面图1

Best Cuts and Colours 美发店是现有品牌，在墨尔本最受欢迎的购物中心又开设了一个沙龙。这个不断扩大的品牌已经拥有了十多年的行业经验，Equator 设计事务所设计的这家店是该品牌的第十五家分店。

设计的整体理念是用极简手法和现代元素诠释时尚、美和品质。沙龙的门面由带有标识的漆面木板构成，这种木板也构成了前台及白色 LED 灯装饰的置物架的框架结构。亚光地砖与光滑的黑白室内装饰形成强烈的对比，增添了几分温暖的气息。此次设计的重点是镜子的运用和天花板上配置的灯具。灯具从天花板悬挂下来，正好对着每一面镜子，让光线自镜面反射开来。每个美发座席的设计都很简单，配有单独的落地镜面，黑色皮革椅子搭配柔和的白色墙面，构成了这样一个极具特色的沙龙空间。

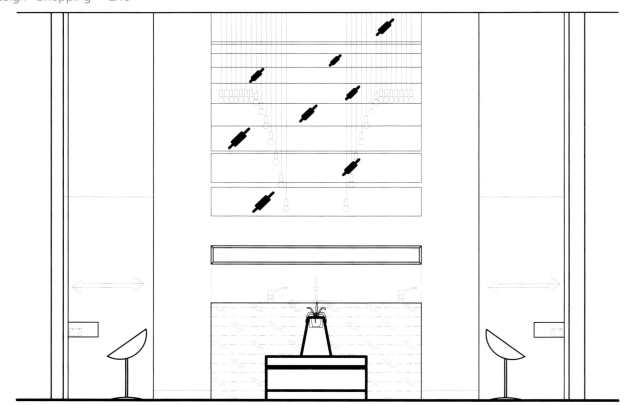

Elevation 2　立面图 2

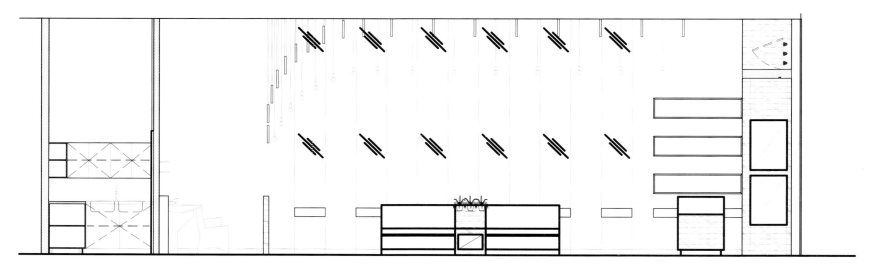

Elevation 3　立面图 3

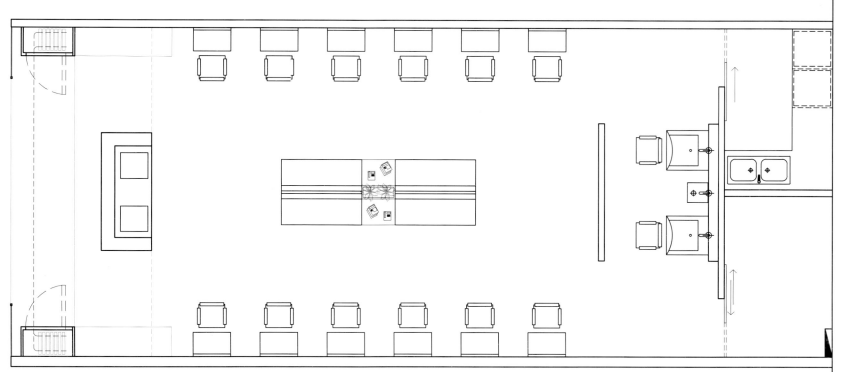

Plan　平面图

M'z Hair Design

M'z Hair Design 美发店

Design agency : Matsuya Art Works Co., Ltd.
Designer : Tetsuya Matsumoto
Location : Japan
Area : 66.2 m²
Photography : Toshiyuki Nishimatsu

设计单位：松屋艺术设计有限公司
设计师：松本哲也
项目地点：日本
项目面积：66.2 平方米
摄影：Toshiyuki Nishimatsu

We placed white panels in arbitrary sizes on two walls and installed mirrored walls on the other two ends of the room. The optical effect of the coupled mirrored wall produces unlimited patterns of the white panels, while creating an imaginary spacious room.

设计师在两侧墙上安装了许多不同尺寸大小的白色木板，而在房间的另外两端则安装了镜子。两面镜子墙与白色面板相互映照，产生了多变的光学效应，从而创造出一个充满想象力的广阔空间。

Elevation 立面图

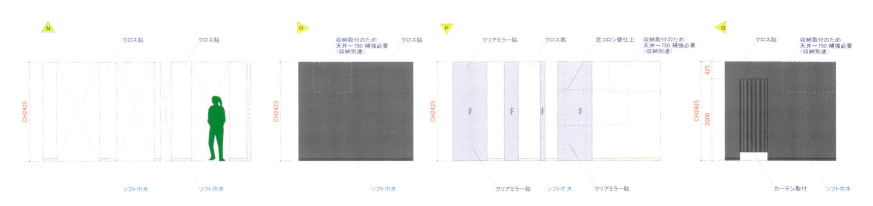

Cutaway View 1 剖面图 1

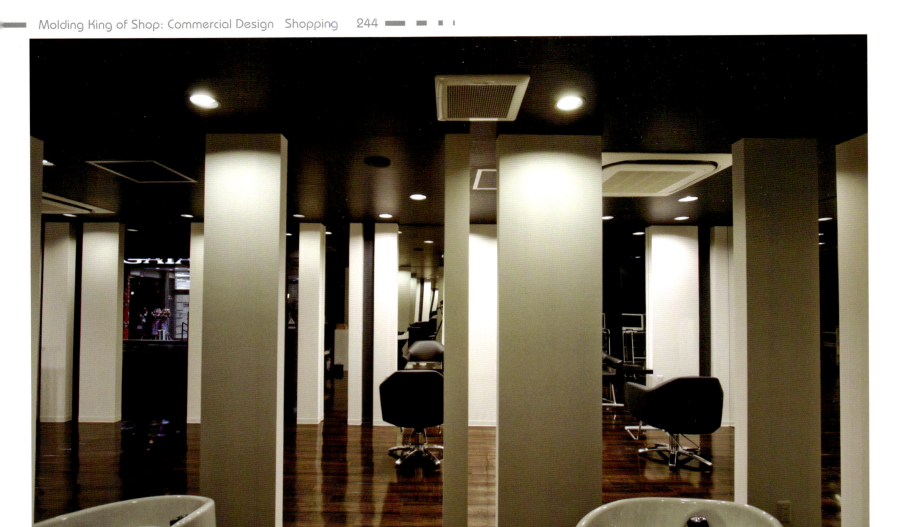

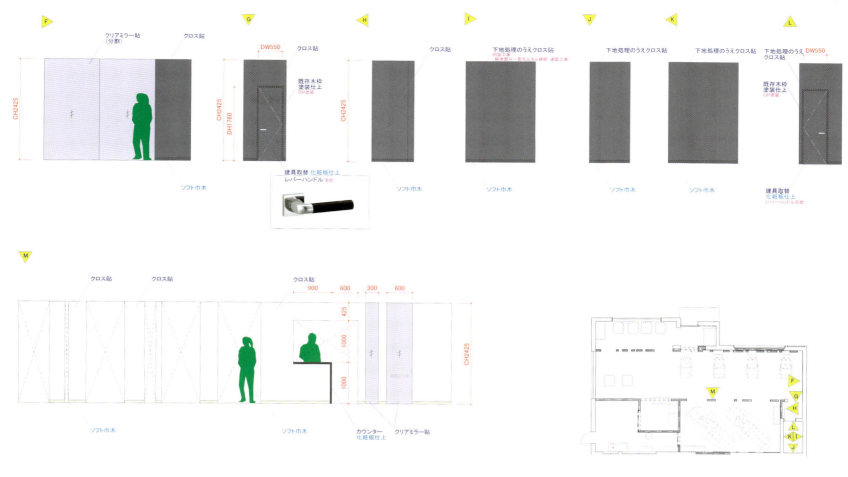

Cutaway View +Plan 1　剖面图 + 平面图 1

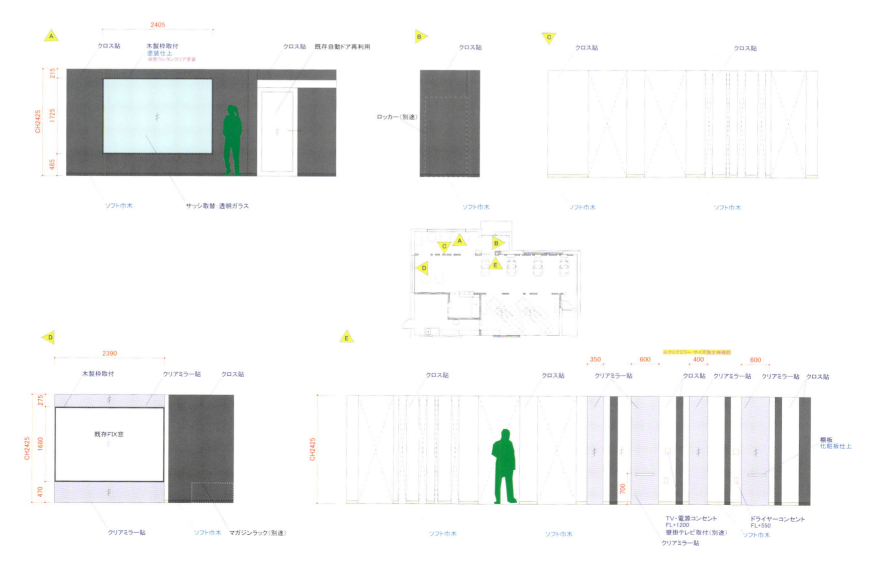

Cutaway View +Plan 2　剖面图 + 平面图 2

满目琳琅迷人眼
Exquisite items or ornaments brightening your eyes

沧海遗珠，是那样珍贵难寻，正如那些隐藏在繁华闹市的华丽商铺，一样让人心动，在喧哗街头给人惊喜。

Just like a piece of precious pearl in the ocean which is hard to be discovered, the luxury stores hidden in the sea of crowd and bustling street are also attractive, making a big surprise to those passers-by around the corner.

Hao Jing Leisure
汕头豪景休闲会所

Design agency : Shantou Youxin Decorative Engineering Co., Ltd.
Designers : Wu Miaochang, Lu Zhonglian, Wu Qiaopeng, Ou Lisong
Location : Shantou, Guangdong, China
Area : 3,600 m²
Photography : Wang Yuxun

设计单位：汕头市友信装饰工程有限公司
设计师：吴淼昌、卢中炼、吴乔鹏、欧利松
项目地点：中国广东汕头
项目面积：3 600 平方米
摄影：王毓洵

This is a large leisure club in Chaoshan area, with entrance hall, café, catering and public bathing room, resting hall and 71 luxury massage rooms, covering a surface of 3,600 m². It promotes a new concept of "VIP Integrated External Suite", which offers a noble, healthy, sunny and leisure new experience. Decorative design stresses on natural, gentle and ecological methods, to deal with the structure according to the environment. An atrium is used to create spacious sense with the cooperation of lights, and thus avoiding the shortage of space-abundant horizontal area and insufficient vertical height in the entrance hall and public bathing area. Materials transition, landscapes changing in different angles, integrated functions all these designs aim to solve the problem of small space caused by the limited space of "VIP Integrated External Suite". Façade is largely made by wood with an array shaped order, while the proper lights seem to say that they don't attract consumers with colorful light, but only to express gentle care.

1F Plan　一楼平面图

该会所是潮汕地区规模较大的一家保健理疗休闲中心，营业面积有3 600平方米，配备有迎宾大堂、咖啡吧、自助餐厅、公共浴区、休息大厅和71间豪华按摩房。于本地首推"VIP集成外置套间"新概念，全程打造尊贵、健康、阳光、休闲新体验。

装饰设计强调自然、生态、温馨的手法，因地制宜地协调处理构造环境。用"天井中庭"的造型配合灯光效果营造竖向延伸感，以此来规避迎宾大堂和公共浴区等水平面积充裕、垂直层高不足的空间瑕疵；利用材质变化、移步换景、集约功能来解决"VIP集成外置套间"体积小巧导致空间局促的问题；外观立面改造使用大量生态木，以阵列的规律形成线性的秩序，点到为止的灯光语言仿佛在诉说："这里的灯光不用光怪陆离来哗众取宠，这里的灯光无需喧嚣，表达的只是温柔的关怀"。

总面积约：2450.00m²

2F Plan　二楼平面图

Qinglan Peninsula Club
清澜半岛会所

Design agency : Shanghai Xiao's Design&Decoration Co., Ltd.
Designer : Ibin Shaw
Location : Wenchang, Hainan, China
Area : 2,000 m²

设计单位：上海萧氏设计装饰有限公司
设计师：萧爱彬
项目地点：中国海南文昌
项目面积：2 000 平方米

Qinglan Peninsula is surrounded by sea from three sides and shares an irresistible seaview. This building is in Southeast Asian style and the interior design is the extension of this style called Orientalism conception. It is formerly used as a sales office, but changed into reception after the houses are sold up. Through different design methods the central hall will be designed as a hall as soon as the existing sand table is moved away.

Seascape facing the south is symmetric with the passage way at the entrance, forming a negotiating area, which is a settle area for those customers after visiting the showroom and know better about the information of those houses. This is the goal of developing this area: deals could be easier with the help of wonderful seascape and softest sofa while drinking tea. Furnishing and decorations are made especially for this space by designers: chairs are the recently rewarded piece of Ibin Shaw, and they can make this place much more elegant. Behavior and selling psychology are all considered in this case, and interpreted in the space. As a result, this design has achieved what we wished.

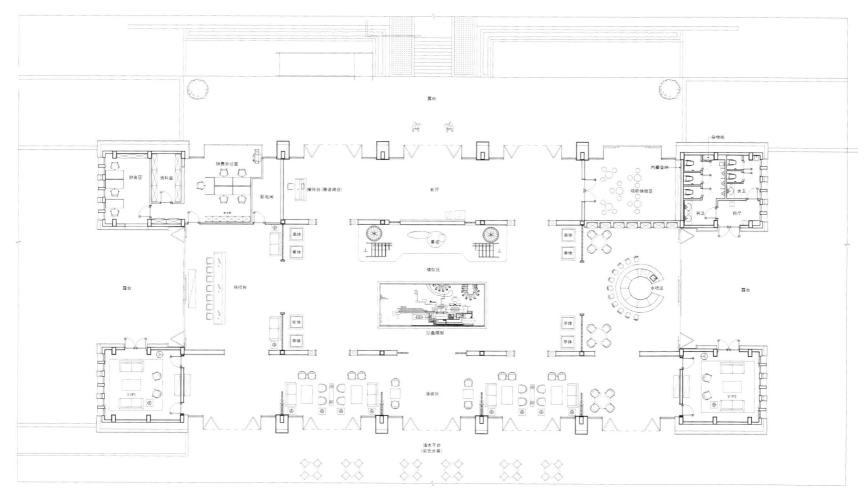

1F Plan 一楼平面图

清澜半岛三面环海，拥有无敌海景。建筑为东南亚风格，室内设计延续此风格，即为东方主义设计理念。会所初期做售楼之用，完成使命后可能要作为纯粹的接待之用，因此采用各种设计方法，现有的沙盘将来撤走后中厅即为大厅。

朝南的一线海景，与入口的长廊彼此对称，成为"洽谈区"。这是客人参观完样板房、看好楼盘信息后的落定之地，也是开发楼盘的目的所在，把最好的位置、优美的海景、最舒适的沙发和座椅给客人，一边欣赏美景，一边品茗，合作自然容易成功。家具和配饰都是设计师为此空间量身定做的，椅子是主设计师刚获奖的作品，用在此地相得益彰。萧氏设计把行为心理和销售心理都完整地体现在该设计空间里，这样的设计一定会达到预期的效果，事实证明确实如此。

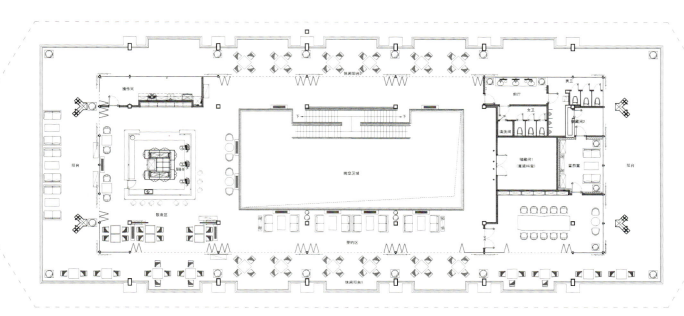

2F Plan 二楼平面图

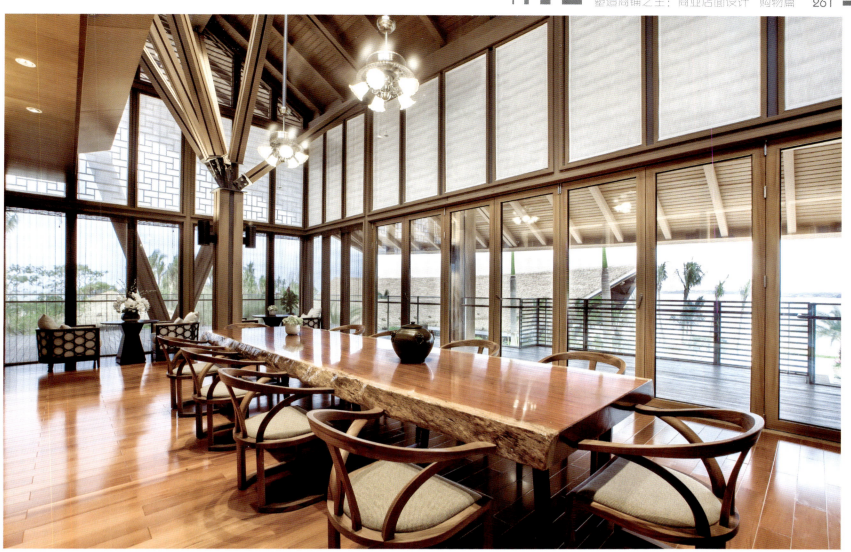

Casa VB Store
Casa VB 生活专卖店

Design agency: David Guerra Architecture
Location: Belo Horizonte, Brazil
Area: 480 m²
Photography: Jomar Braganca

设计单位：David Guerra 建筑事务所
项目地点：巴西贝洛奥里藏特
项目面积：480 平方米
摄影：Jomar Braganca

In the remodeling of Casa VB Store in the city of Belo Horizonte, Brazil– which sells curtains, tapestry, fabric and other items for home interiors – blue metal blinds were used to make the façade more modern, protecting from excessive lighting while maintaining natural ventilation. One of the premises was to create more visual unity in a store that sells a wide range of products with their own singularity, which led to making the interior lighter and clearer by the use of white tones, and to using individual lighting to bring evidence to each product. The high ceiling, at the end of the store, integrates the three floors. The interior was designed as a showcase of the products, allowing customers to get a better feel of what is possible to create with them. Art pieces give a contemporary and warm touch to the space.

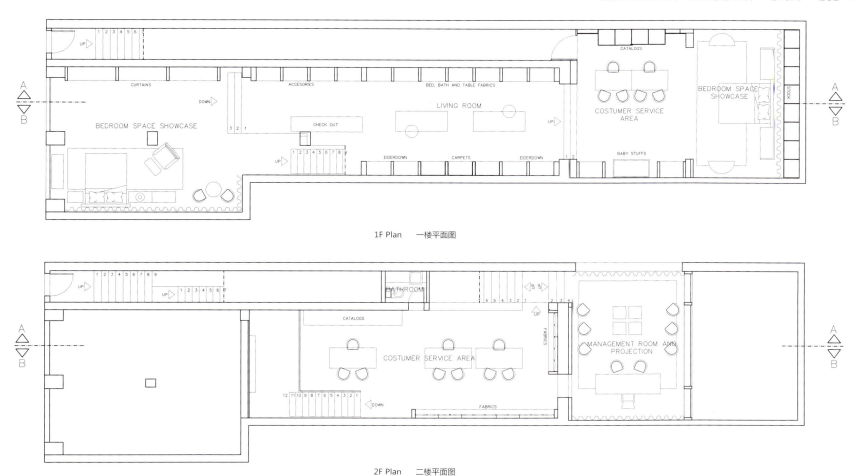

1F Plan 一楼平面图

2F Plan 二楼平面图

该店位于巴西贝洛奥里藏特市，专卖窗帘、挂毯、纺织品等家居室内用品，因此设计采用了蓝色金属百叶窗装饰外立面，不仅让店面更具现代感，还能保证自然通风并且防止过度照明。设计的一个前提就是在店内创建多个视觉整体，以展示颇具特性的一系列产品。通过白色色调的运用将空间衬托得更为清爽干净，并且对每个产品都提供了有针对性的照明设计。店铺尽头的天花板整合了三层空间，内部的设计就如同一个产品的展示柜，让顾客亲身体验产品的创造性。而其间的艺术品则赋予空间现代温暖的格调。

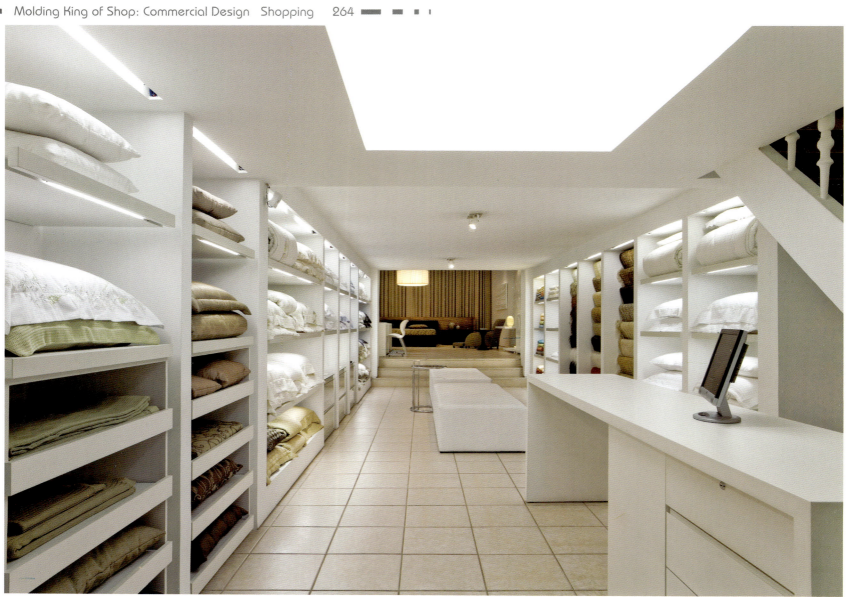

MUJI to GO Shop in Venice

威尼斯无印良品

Design agency : Roberto Murgia Architect, Aliverti Samsa Architect
Location : Venice, Italy
Area : 90 m²
Photography : Giovanna Silva

设计单位：罗伯特·莫吉亚工作室、艾丽弗缇·萨姆沙建筑事务所
项目地点：意大利威尼斯
项目面积：90 平方米
摄影：乔凡娜·席尔瓦

Every new Muji store is essential, being made for displaying the products. It reflects the brand's packaging that is reduced to the minimum, just like a showcase without showing off. A new project for each city but all linked together. Different outlets related to a single philosophy. The philosophy of simplification: reducing any unnecessary gesture, decoration and excess. The store is on two levels, a small room at street level and the rest at the first floor. On the ground floor, in a long and narrow space, glass shelves are used to accommodate merchandise. Light colors, natural materials and matte surfaces are used there. White is used throughout, with some stainless steel and timber exceptions.

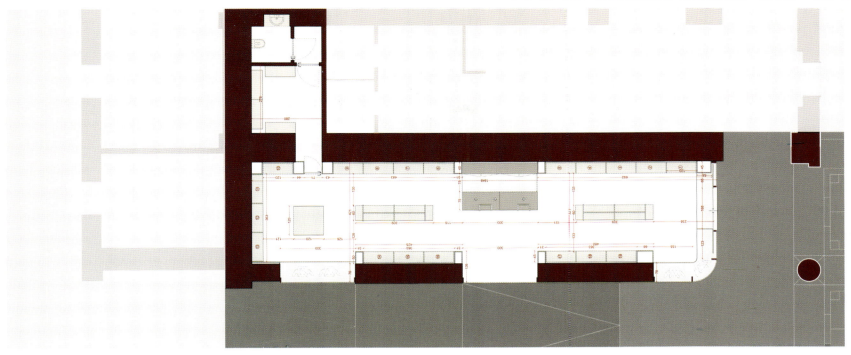

Plan 平面图

尽管只是用来展示商品，但每一家新开的无印良品店都是精品之作。通过最简洁的包装来凸显品牌的魅力，就如同一个低调的橱窗在静静地展示着产品。各个城市的无印良品店彼此之间都息息相关，以不同的方式演绎着同一种风情。无印良品的宗旨就是简洁，没有多余的装饰。该店面共有两层，位于临街铺位。一楼是一个狭长的空间，摆放了陈列商品的玻璃货架。同时使用了浅色自然的材料，表面则是亚光的。白色贯穿整个空间，空间中穿插着不锈钢和木材制品。

Doctor Manzana

Doctor Manzana 品牌空间

Design agency: Masquespacio
Designer: Ana Milena Hernández Palacios
Location : Valencia, Spain
Photography: David Rodríguez

设计单位：Masquespacio 工作室
设计师：安娜•米莱娜•埃尔南德斯•帕拉西奥斯
项目地点：西班牙瓦伦西亚
摄影：大卫•罗德里格斯

Masquespacio presents their last project realized for Doctor Manzana, a store specialized in technical service for smartphones and tablets, besides being a seller of design gadgets for mobile devices. The project consists in the redesign of Doctor Manzana's branding and the realization of the design for their first point of sale located in Valencia, Spain.

The colors were inspired by the name of the brand, which includes "doctor", so they based the concept on a hospital. Eliminating a conventional design, what was left was the color scheme of blues and greens. The blue curtains are a nod to hospitals. Galvanized metal is used to bring a bit of an industrial look to the space.

The striking identity begins with the façade, which incorporates the same angles and the blue and green colors of a doctor's scrubs, along with mixing in, "the salmon color for the fashionistas and the purple for the freaks." A technological air blows through the store, while some details like the blue curtain refer in a metaphorical way to a hospital. Materials like the galvanized steel sheets are doing their more industrial work in the space, while white furniture is offering a light warm touch to the whole.

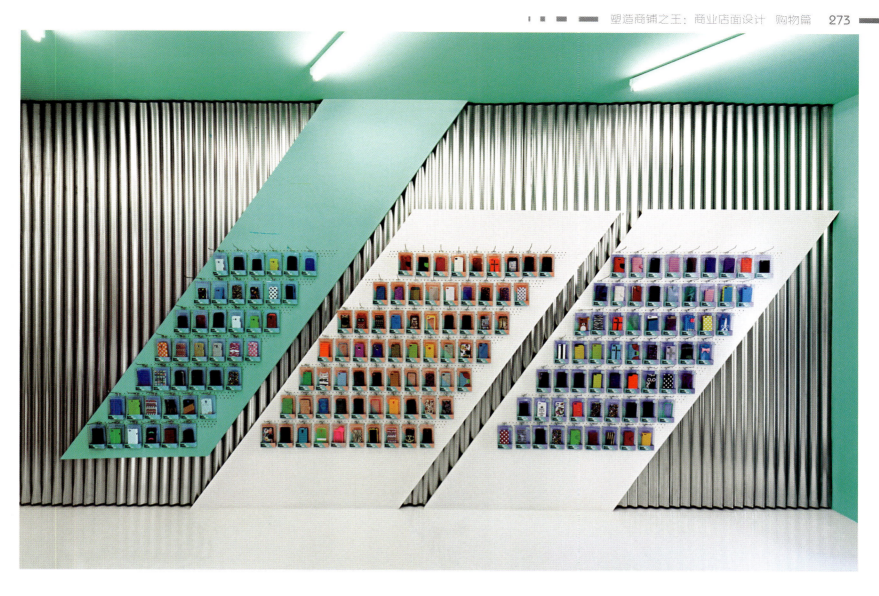

西班牙 Masquespacio 工作室展示了其为 Doctor Manzana 设计的最新作品。Doctor Manzana 为智能手机和平板电脑提供技术服务以及销售手机配件的品牌商（Manzana 西班牙语意为苹果）。本案包括重新设计品牌形象，以及为位于西班牙瓦伦西亚的第一家实体店进行设计。

整个店铺的色彩搭配、墙面的处理、天花板上的照明都体现着科技感。本案色彩的搭配来自品牌的名称，其中包括"医生"，所以设计理念基于医院，摒弃传统的设计，只保留了蓝色与绿色的配色方案，并运用镀锌的金属管使空间散发出一些"工业化"的设计风格。

外立面的设计极其醒目，其延续室内的配色方案，在大面积蓝色和绿色下点缀着时尚的橙黄色和奇异的紫色。整个店铺散发着科技感，一些细节之处，如蓝色的窗帘以一种隐喻的方式与医院联系在一起。镀锌钢板等材料的运用使空间更具工业感，白色家具则带来一丝温暖的感觉。

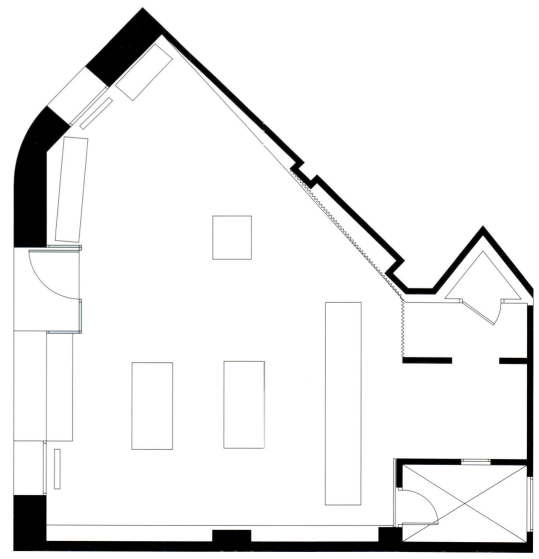

Plan　平面图

Whites Dispensary
白色药房

Design agency : Studio Equator
Designers : Carlos Flores, Benjamin Fretard
Location : Melbourne, Australia
Area : 180 m²

设计单位：Equator 设计事务所
设计师：卡洛斯·弗洛雷斯、本杰明·弗勒塔尔
项目地点：澳大利亚墨尔本
项目面积：180 平方米

In our design for Whites Dispensary, Studio Equator wanted to achieve a distinctly different look in an industry that is deeply rooted in a traditional and unchanging aesthetic. The aim was to design an interior and visual identity to increase the dispensary income whilst maintaining steady growth in the beauty, health and luxury goods sections, to present through an environment that it emphasizes personal service and is a trustworthy brand.

The ultimate goal is to reposition the brand to better service the market, and in doing so, change the way companies behave in an industry that hasn't changed in decades. With this refreshed concept, Studio Equator has allowed the visual identity to grow with the ability to change quickly and keep up with customer demands and challenges. Reclaimed wooden horizontal panelling cloaks the façade of the shop, and showcases the new logo. An organic use of materials and authentic chemist and medicinal symbols, support the overall design.

Colour, materials and a finishes scheme of raw timber, green, white, black and gold separate each section – Health, Wellbeing, and Beauty. The interiors reflect a contemporary look & feel often showcased in the fashion and hospitality sectors. The checkerboard tiled floor, custom joinery, reclaimed grey skin aged timber, decorative concrete bricks, green carpet, cross-shaped large scale lights and industrial furniture create a sense of warmth framing the products and allowing them to stand out. The beauty products contrast with the natural feeling of the dispensary proposition and thus stand out in its earthy luxurious approach.

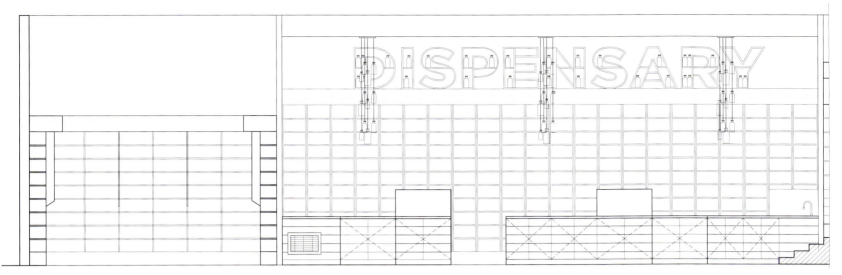

Elevation 1 立面图 1

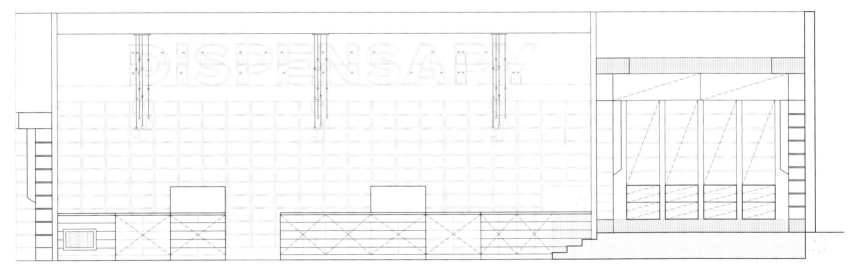

Elevation 2 立面图 2

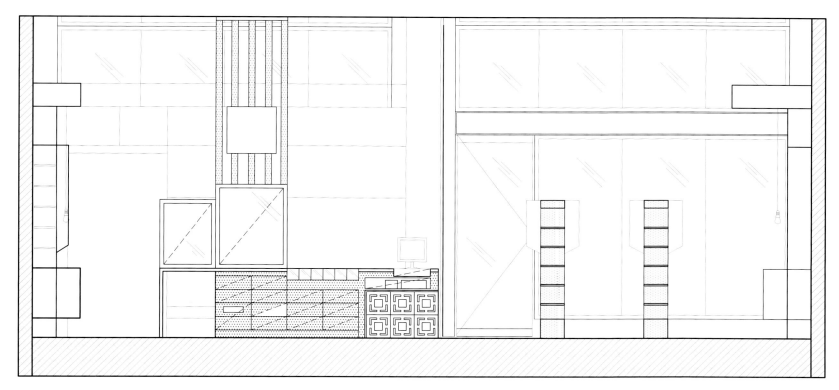

Elevation 3　立面图 3

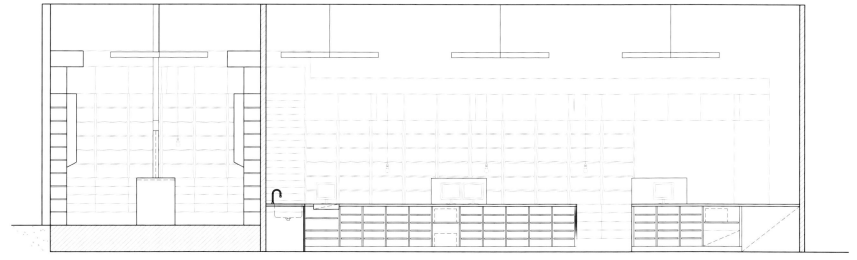

Elevation 4　立面图 4

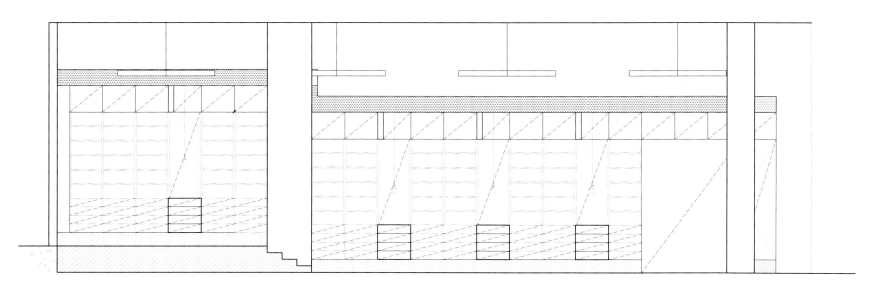

Elevation 5　立面图 5

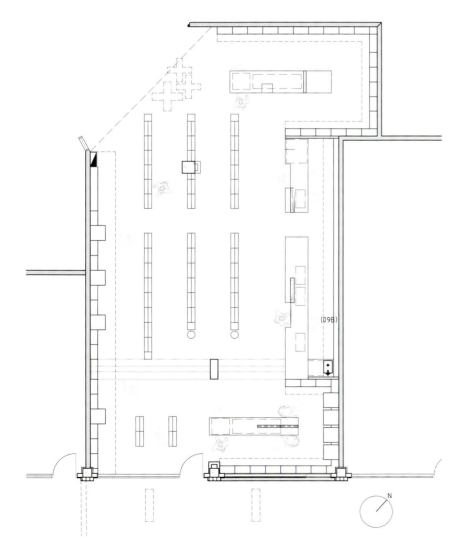

Plan 平面图

在白色药房的设计中，Equator 设计事务所想设计出一个该行业的全新形象，它将与根深蒂固的传统美学观念大相径庭。设计师旨在通过室内设计，以及对药房的视觉识别进行设计，以增加药房收入，促进美容品、保健品和奢侈品销售稳定增长，通过环境向外界展示其重视个体服务的形象，表明它是个值得信赖的品牌。

重新定位品牌形象，为市场提供更好的服务，从而改变数十年来行业的整体行为方式，这就是本案设计的终极目标。抱着这样一个全新的想法，Equator 设计事务所提升了药房的视觉辨识度，使它快速转变，与顾客的需求同步，迎接挑战。木制镶板水平搭建，制成药房的门面，然后再展示全新的商标。店内材料的有机使用，另外配备专业的药剂师和药物标志，对药房的整体设计起到了支撑作用。

设计采用木材为原料，涂上绿、白、黑、金黄等颜色，将卫生、保健和美容等区域区分开来。时装店和酒店就常常采用这种室内设计方法，以体现当前的时尚潮流。药房内铺上棋盘式花纹地板砖和绿色的地毯，安装十字形大型吊灯，放上定制的细木工制品、翻新的灰色古木器具和一些工业家具，营造出一种温暖的氛围，烘托店内的产品，使它们更加引人注目。美容产品与药房主题的自然气息形成对比，在朴实而又奢华的设计背景下格外显眼。

Masters Craft Palace Hotel Tokyo
巧匠工艺品店

Design agency : Love the Life
Designers : Akemi Katsuno , Takashi Yagi
Location : Chiyoda-ku, Tokyo, Japan
Photography : Shinichi Sato

设计单位：Love the Life 工作室
设计师：勝野明美、八木孝司
项目地点：日本东京千代田区
摄影：佐藤真一

This store is in the long-established hotel at Otemachi rebuilt in 2012. South side of the building is facing the Wadakura Funsui Kohen park. This area has a valuable view of openness in the business area of the downtown. However, this store almost has nothing to do with the good location because it is in the basement of the building.

Before beginning to design, we were able to visit the Masters Craft headquarters in Mizunami-shi, Gifu and the neighboring wide areas. There are deep green mountains and courses of river, and sunny settlements appear intermittently. The state gave a mixed sense of nostalgia and freshness to us live in urban areas. If that temperate forest is reproduced allegorically in this small place close to the Imperial Palace in some fashion, it will not be meaningless.

First, we carefully designed wooden shelves covering two vertical plane at the back of the store. Those frames of sharp angle are illuminated by indirect lighting. This simple pattern symbolizes the appearance of woods and mountains of the Tono region. 14 poles are fixed to the ceiling in the middle of the store. These squared lumber has faint shade of gray tones and emphasize depth and height of the store. Strip-shaped tiles cover the wall behind the cashier casually represent the origin of Masters Craft, a manufacturer of ceramics.

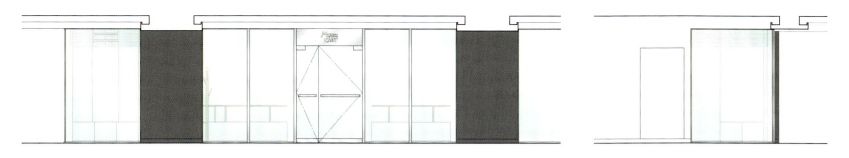

Elevation 1　立面图 1

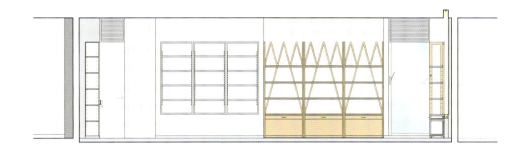 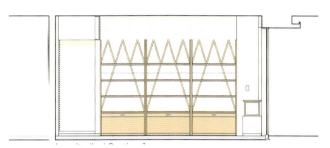

Elevation 2　立面图 2

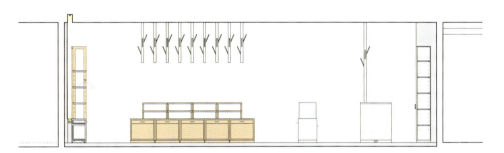 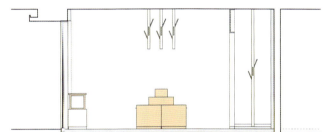

Elevation 3　立面图 3

重建于 2012 年的 Otemachi 酒店历史悠久，建筑南侧面临和田仓喷泉公园，这使得该区域在繁华的商业区中拥有极佳的开阔的视野。然而店面却和良好的地理条件毫无关联，因为它位于酒店建筑的地下部分。

在开始构思之前，设计师拜访了位于岐阜县瑞浪市的总部以及周边地区。密林、深山、河流和明媚的阳光给久居城市的设计师带来灵感，他们决定把温暖的森林气息带到这个坐落在皇宫附近的小店中。

首先，设计师在店内两个墙面设置木货架，这些受间接光线照射的木架仿佛远山和树林。天花的中心位置吊装了 14 个颜色深浅渐变的木制装置。收银台背墙以鲜艳的条形瓷砖覆盖，使人联想起巧匠工艺品店作为陶瓷制造业的渊源。

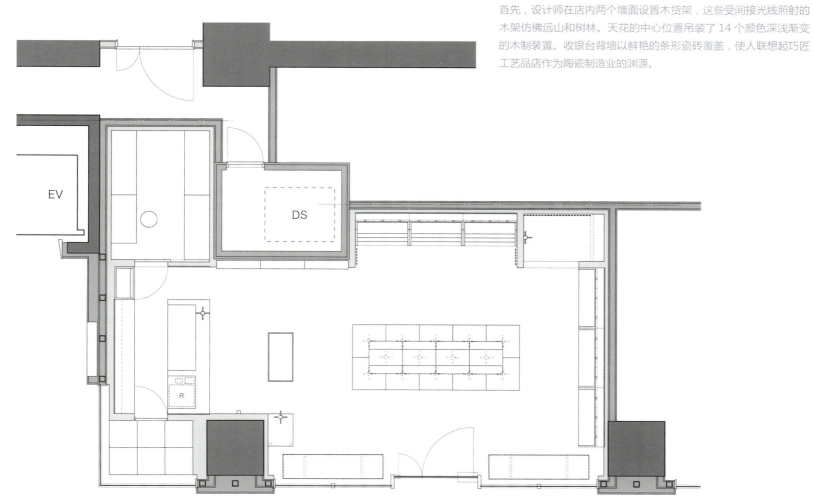

Plan 平面图

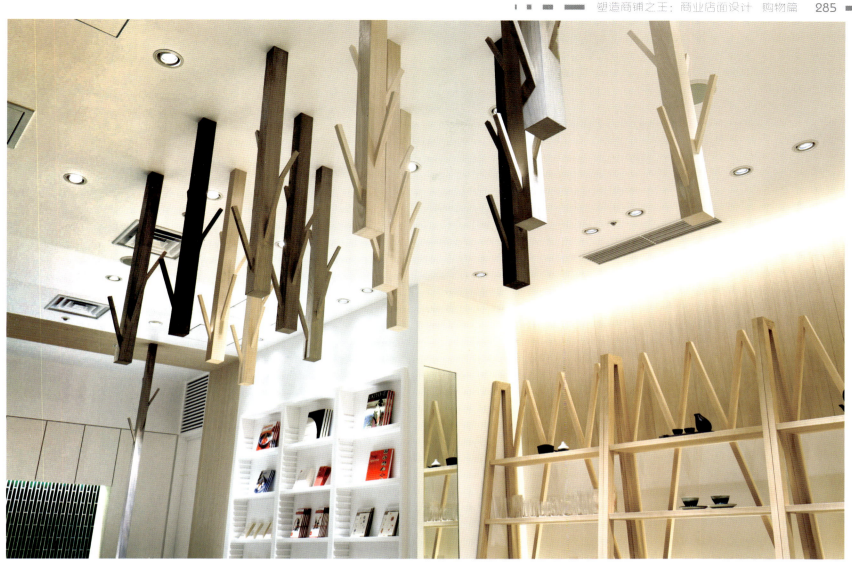
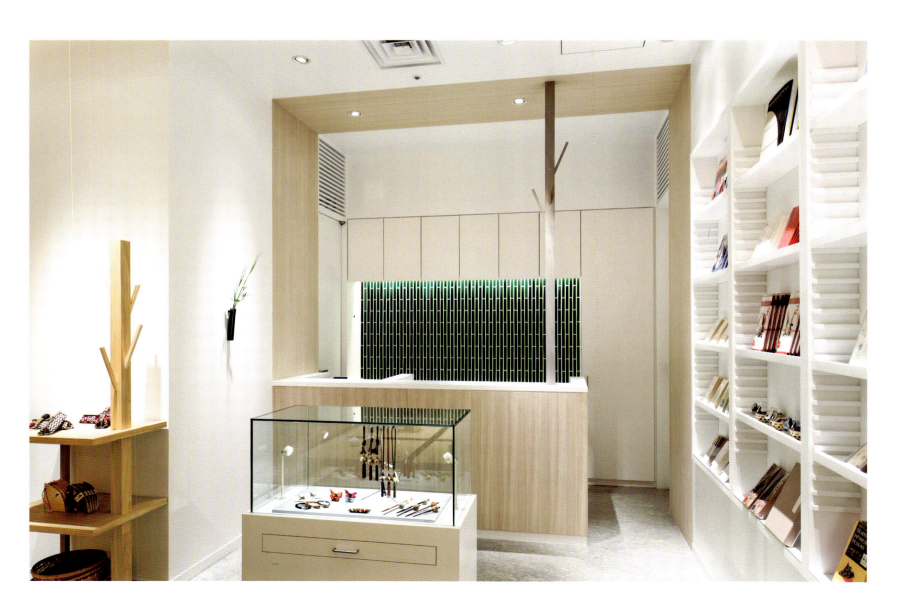

Aleybo Flagship Store

Aleybo 厨具用品旗舰店

Design agency : 5 Star Plus Retail Design
Designers : Bruce Li, Barbara Seidelmann, Ray Zhang
Location : Beijing, China
Area : 100 m²
Photography : Bruce Li

设计单位：斐思达品牌设计咨询有限公司
设计师：Bruce Li、Barbara Seidelmann、Ray Zhang
项目地点：中国北京
项目面积：100 平方米
摄影：Bruce Li

Aleybo is a brand that offers a range of European and American kitchen and lifestyle accessories. Unique and innovative design, functionality and the impact of color are the product characteristics that play an essential role for the brand.

Based on the dynamic brand image and requirement of using the space in different ways, a flexible solution for the window area was needed. Since the original store design had not planned for a window display and store overall construction works had only started, the designers had a free card to come up with a creative solution. The goal was therefore to create something smart, elegant, and simple, that would support the rest of the store design yet create a big impact and draw people into the store.

For a maximum impact, the designers built on the strong, colorful design of the products and designed large posters, which are hung from rails behind the shelves. The position of the posters can be changed. Most importantly, different store window display scenarios can be installed with a varying number of posters in different shapes, sizes and locations based on seasonal and marketing needs. During events, the store receives a sense of privacy with a semi-translucent curtain hung on the rails where usually the posters are placed.

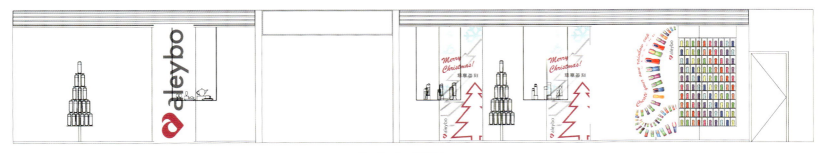

Elevation 1　立面图 1

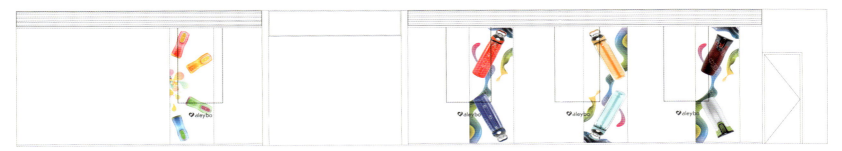

Elevation 2　立面图 2

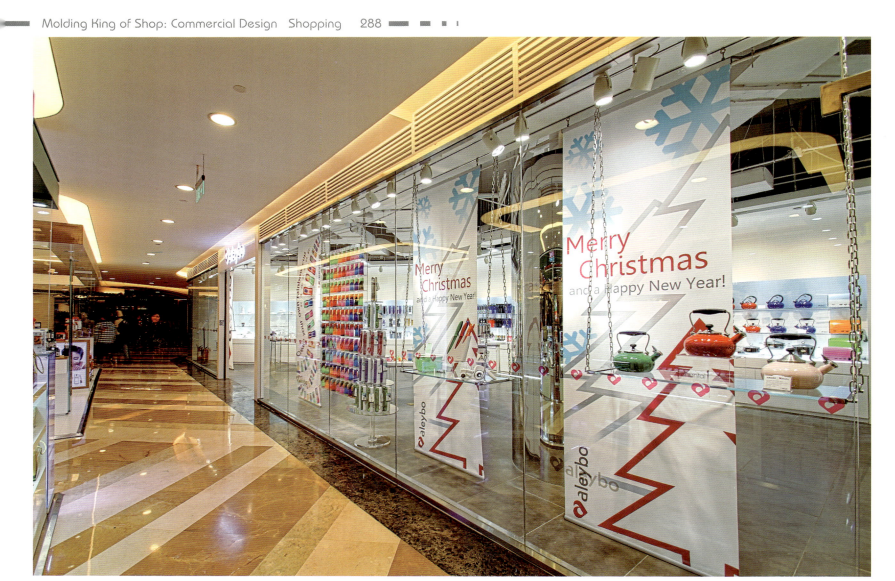

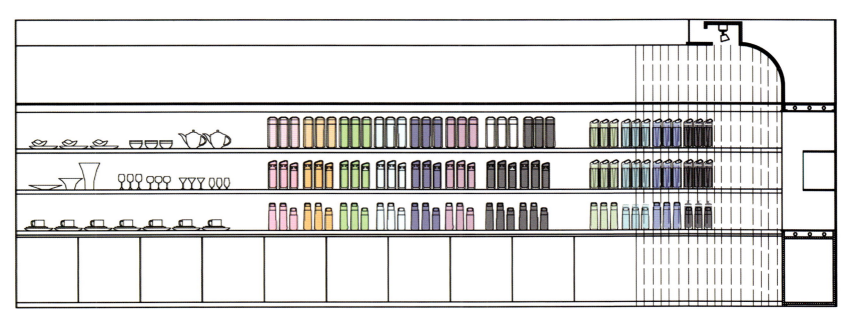

Elevation 3　立面图3

Aleybo 是一家销售欧美高级品牌餐厨及生活用品器具的国际知名零售商。独特而充满创意的设计风格、功能性以及色彩冲击力是这个品牌所销售的产品最重要的特色。

考虑到其充满活力的品牌形象和对空间格局大胆分割的特点，需要设计一种更灵活，并具有可调整性的橱窗解决方案。由于最初的整体店面设计方案中没有包括具体的橱窗设计这一部分，并且当时整体装修施工才刚刚开始，故而有相当的自由度来考虑用一种更有创意的设计思路进行橱窗设计。设计的目标是让橱窗陈列展示出优雅、简约的风格来配合店内其他部分的设计，给人留下深刻的印象，并最终吸引顾客进入商店。

为了最大限度地提高视觉冲击力，设计师还设计了重点突出、色彩丰富的产品大海报，将其悬挂在商品展示货架的后面，可以自由移动和变换位置。最为重要的是，根据不同的季节或者市场需求，商店橱窗展示主题可以通过多种不同形式、数量、尺寸、位置的海报展现出来。此外，在特定店内活动期间，商店也可以通过在海报的位置悬挂半透明隔帘的形式起到隔离和保护隐私的作用。

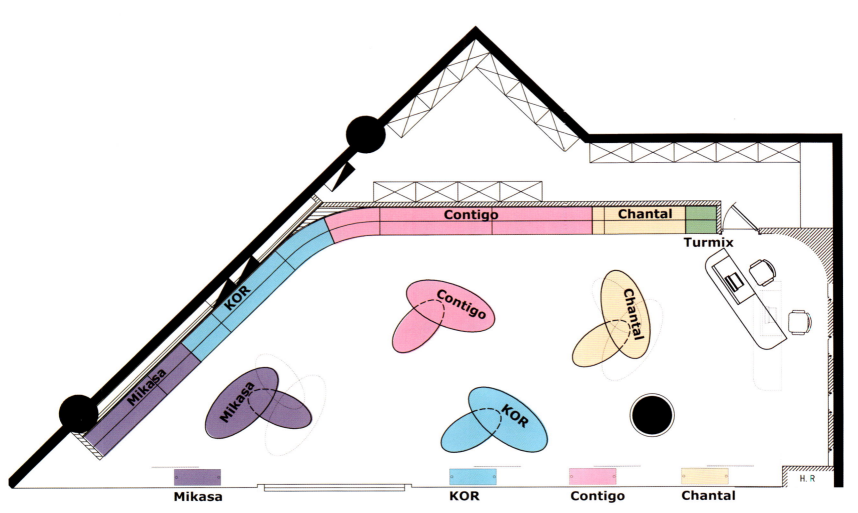

Plan　平面图

Farm Direct

Farm Direct 概念店

Design agency : PplusP Designers Ltd
Designer : Wesley Liu
Location : Hong Kong, China

设计单位：维斯林室内建筑设计有限公司
设计师：廖奕权
项目地点：中国香港

This is Farm Direct's first concept store. The store's principal business is in the sales of 100% Hong Kong locally produced hydroponic vegetables. The design concept is about using the cheapest and the most primitive materials to create a European styled store. The red bricks forming the bright exterior are overlaid with a brushed white finish along with a sprayed green vignette, which is a reflection of the fresh produce available in store. The interior floor is laid with volcanic rocks, a characteristic used in this specialized method of farming. Linen textured carpet lining the wall creates a space of warmth and coziness, whilst the unfinished wood panels, that are used to construct the display cabinets, continue the fresh and natural theme. The visualization of vegetable roots and stems is achieved by the handcrafted LED ceiling-hung light fixtures.

本案是 Farm Direct 的首家概念店。该店主要出售香港本土出产的水培蔬菜。设计理念是用最原始、最便宜的材料打造出欧式风格的店铺。非常耀眼的外墙用一般的红毛砖砌成，刷上白色油漆，再喷上绿色的装饰图案，传达出店内出售新鲜蔬菜的信息。 室内地面用火山石铺就，墙身贴上具有麻布质感的地毯，为空间营造出温馨、舒适之感。柜身用天然木条拼接而成，延续清新自然的主题。天花板上垂下的自制 LED 吊灯，感觉就像蔬菜的根部，也为店内增添了独特的视觉效果。

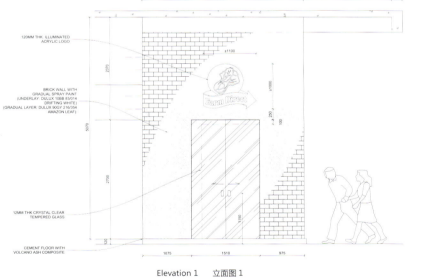

Elevation 1　立面图1

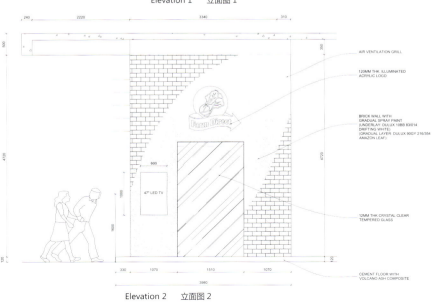

Elevation 2　立面图2

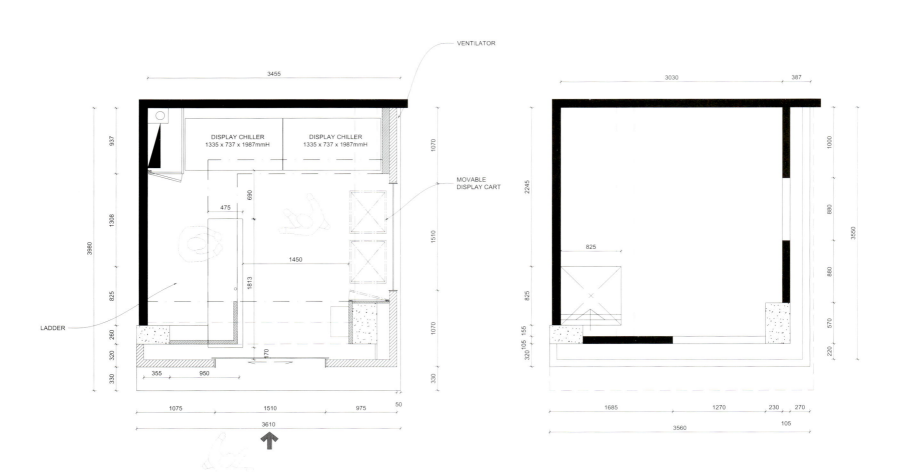

1F Plan　一楼平面图

2F Plan　二楼平面图

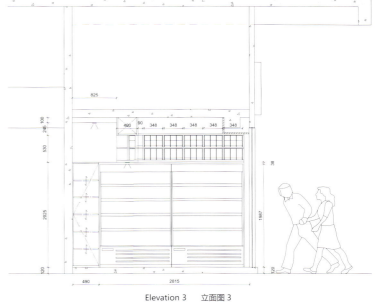

Elevation 3　立面图 3

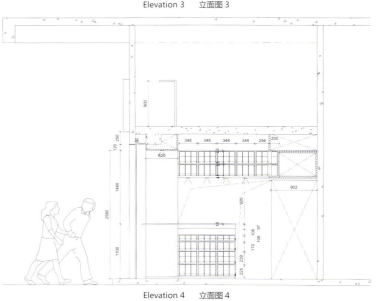

Elevation 4　立面图 4

Retail Design Kabel Deutschland

Kabel Deutschland 电缆零售店

Design agency : Hartmannvonsiebenthal GmbH
Location : Berlin, Germany
Area : 45 m²
Photography : André Müller

设计单位：Hartmannvonsiebenthal 设计公司
项目地点：德国柏林
项目面积：45 平方米
摄影：安德烈·米勒

In order to achieve the same with Kabel Deutschland's shops, design and fitting concept have been revised completely. What strikes the eye is the reduced approach: strict confinement to the primary CI-colors yellow and white, product-centered zoning, functional furniture layout. The result: welcoming Kabel Deutschland shops that are recognized at first sight. The design language with elliptical elements in graphic design and architecture is consistently applied to shop front design as well: elliptical elements are designed for each shop individually in order to adapt the overall design concept to each shop in the retail network.

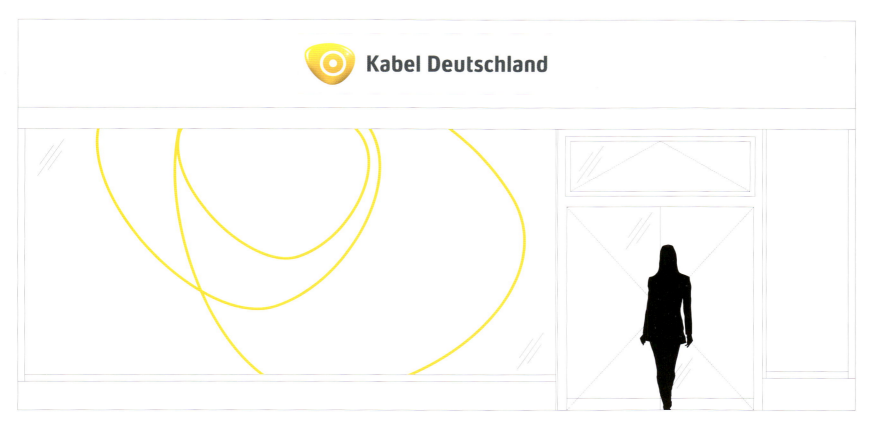

Elevation　立面图

为了与 Kabel Deutschland 连锁店铺取得一致的效果,设计师对该店的设计概念进行了彻底的修改。店内最吸引眼球的设计是黄白搭配的色彩,明确地标示出产品中心区和功能区。这样做的目的就是让人能够一眼认出这是 Kabel Deutschland 零售店的设计。椭圆图案和形状是内部设计和门面设计都用到的元素,每一家零售店都采用椭圆元素设计出各自的特色,从而形成了该品牌的整体设计概念。

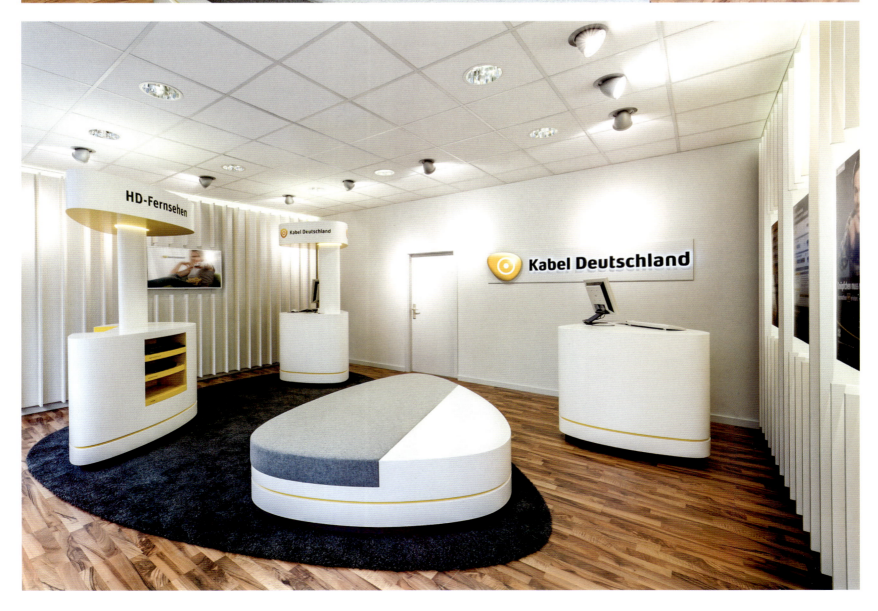

SPAR Flagship Store
SPAR 旗舰店

Design agency : LAB5 Architects
Designers : Linda Erdélyi, András Dobos, Balázs Korényi,
         Virág Anna Gáspár
Location : Budapest, Hungary
Area : 2,000 m²
Photography : Zsolt Batár

设计单位：LAB5 建筑事务所
设计师：Linda Erdélyi、András Dobos、Balázs Korényi、
     Virág Anna Gáspár
项目地点：匈牙利布达佩斯
项目面积：2 000 平方米
摄影：Zsolt Batár

SPAR had the idea of building up a unique interior, which provides a high quality costumer experience. The ceiling is guiding and attracting you from the entrance to the back zone, and then shows different alternative ways to go on. Due to the condition of the modest internal height, we wanted to gain the space above the suspended ceiling zone, so we didn't put a ceiling, unless it was really necessary, and where we put it, it was used in a free-form way, for being presented as an individual, expanded statuesque object. Where we could we used solid white surface, and where we had to put additional elements (lights, sprinkler, etc.) we used optical ceiling.

Because of different use, there are two zones where the optical part of the ceiling converts into a 3d form by flowing down to the ground. At the bakery products' warm feelings are strengthened. At the wine section, the lamellas of the ceiling are continuing down to the ground to form a space of a cellar, and to indicate at this point the quality and the culture of the product. Generally saying, as the ceiling is the element that can be seen from everywhere, it becomes one of the main elements of orientation and impression. Shelves and counters are forming rounded islands together, just as if they were standing at a market.

Acryl (corian) was chosen for the finishing of all rounded furniture, as they had to be white, shiny, clean, durable, and supporting the "fluid" effect. Due to many contradictory specifications we couldn't apply concrete for the floor as we planned, but the single colour solution of grey tiling is perfect for the goal.

Elevation 1　立面图 1

Elevation 2　立面图 2

SPAR 旗舰店希望打造出一个别具一格的室内空间，为客人提供高品质的购物体验。屋顶的设计将引导和吸引顾客从入口处进入商店，随后会有多个不同方向的通道以方便购物。鉴于室内高度的限制，为了扩大使用空间，店面内几乎没有采用吊顶设计，在那些必须使用吊顶的地方则采用了自由的排列方式，使其以一个独立的、展开的、轮廓清晰的方式展示出来。在这些地方，设计师采用了结实的白色表面，在那些必须增加元素的地方，如灯、洒水器等，则采用了透光天花板。

不同货物区的天花板设计也不同；有两个区域的部分天花板延伸到地面形成 3D 效果。

在面包区，增强了产品的温暖气息；红酒区的薄片天花板一直从上延伸至地面处，并形成一个小型的酒窖空间，用以放置一些酒品，并表明此处产品的品位和文化底蕴；这种无处不在的屋顶设计使其也成为整个店铺的主要元素之一，给人留下了深刻的印象。货架和柜台采用了圆岛式设计，仿佛立于市场中。

所有圆形家具的饰面都采用了亚克力大理石，白色、闪亮、洁净、耐用的特点都增强了室内设计的"灵动性"。鉴于一些设计限制，设计师没能采用混凝土地面，但最终选用的灰色瓷砖也非常符合该案的设计理念。

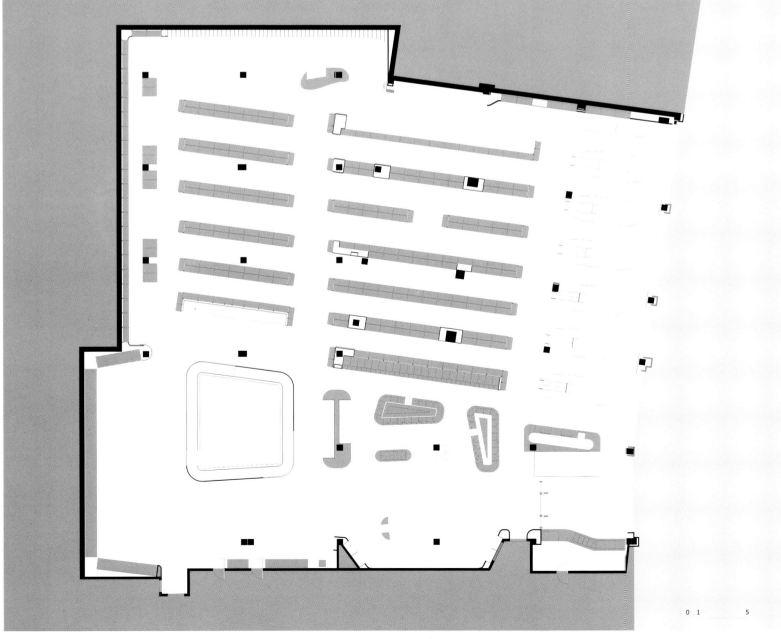

Plan　平面图

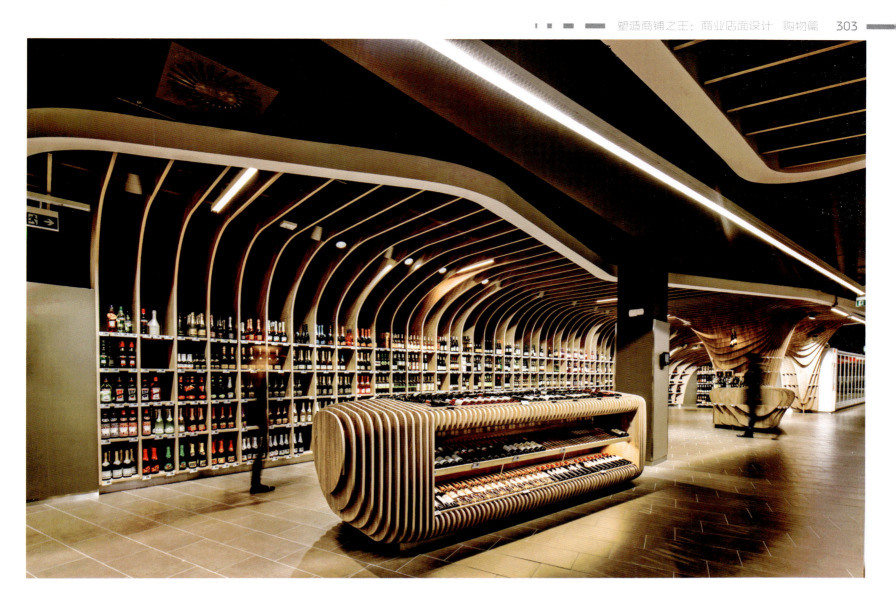

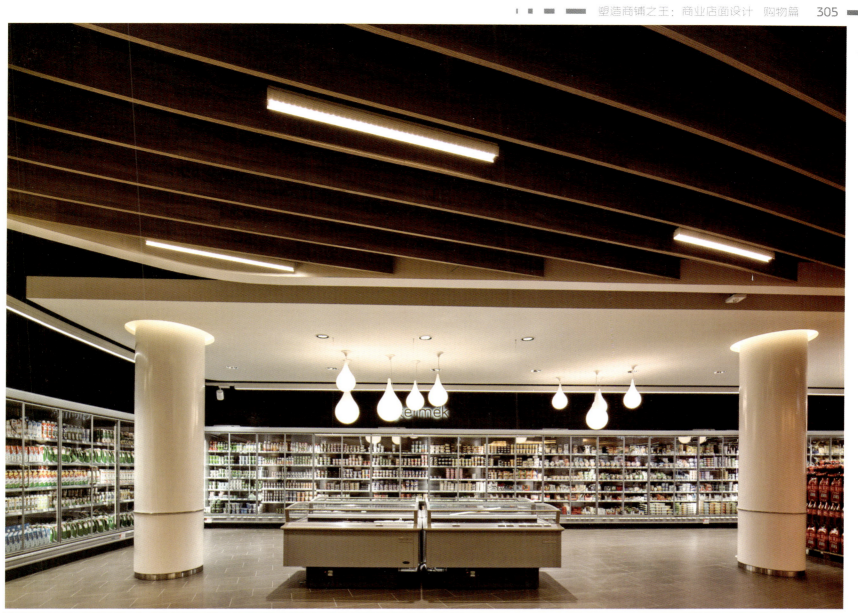

McIntosh AV Galleria
麦景图影音艺廊

Design agency : PplusP Designers Limited
Designer : Wesley Liu
Location : Shanghai, China
Area : 150 m²

设计单位：维斯林室内建筑设计有限公司
设计师：廖奕权
项目地点：中国上海
项目面积：150 平方米

Tasked with the audio system and hi-fi brand's store interiors was Hong Kong-based PplusP Designers. Since their machines are for an exclusive, high-end market, the materials used for the interiors have to align with their products. Black stainless steel represents the feeling of boldness and mechanism. Marble floors are incorporated at the entrance to enhance the feeling of luxury. The store's door handles take inspiration from McIntosh's signature machine, offering a clear brand identity upon entrance.

Passersby will not miss what the designer calls the store's 'time tree', a striking spiral feature with years imprinted at different levels to signify the company's milestones. Lead interior designer Wesley Liu created this spiral feature in order to hide the pillar as well as to create an eye-catching feature for

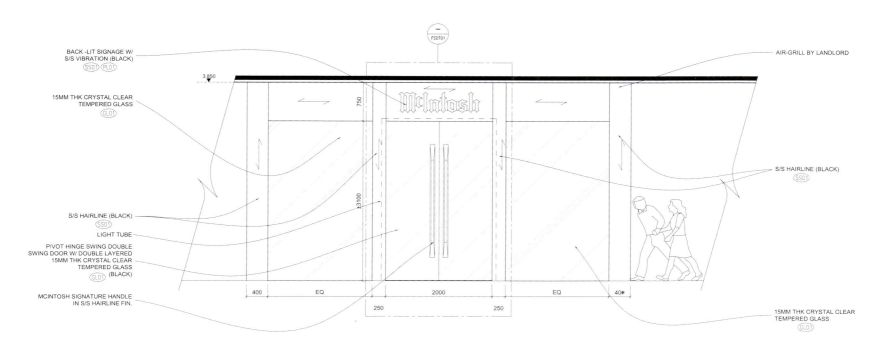

Elevation 1　立面图 1

the shop front to attract customers. It is carefully divided into proportion and features the significant years of achievement. The loose display units come in two forms – one is inspired by the oversized scroll button on the McIntosh machine; the other is representative of a typical hi-fi needle, which is a long Corian unit with crystal glass top.

There are two theatres – a main theatre and a mini theatre. The main theatre features McIntosh Machines' signature and most expansive 2K sound system, while the mini theatre displays a sleek home theatre system. Compared to the main theatre, the mini theatre is more flexible and open.

音响系统与高保真视听器材品牌麦景图影音找来香港维斯林室内建筑设计有限公司负责新店的室内设计。由于该品牌的音响器材专为高档市场设计，室内采用的物料需要配合产品：黑色的不锈钢突出鲜明与机械的感觉，入口的云石地板增添奢华感，而店门把手的设计灵感则源自麦景图影音的招牌产品，为入口营造出鲜明的品牌形象。

路人经过店外，一定不会错过店里那引人注目的"时间树"，螺旋结构上层层不同的年份记录着麦景图影音多年发展历程。这个螺旋结构设计除了要巧妙地隐藏柱子外，亦特意在店内创造惹人注目的元素来吸引顾客。螺旋结构依比例细致分层，每层标注着公司取得重大成就的年份。店内的产品陈

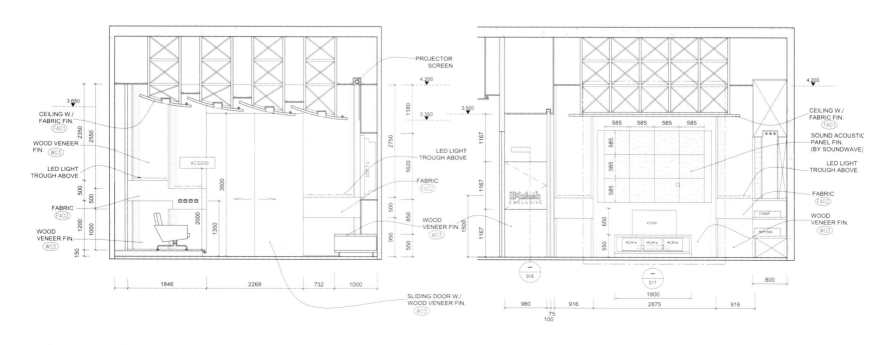

Elevation 2　立面图 2

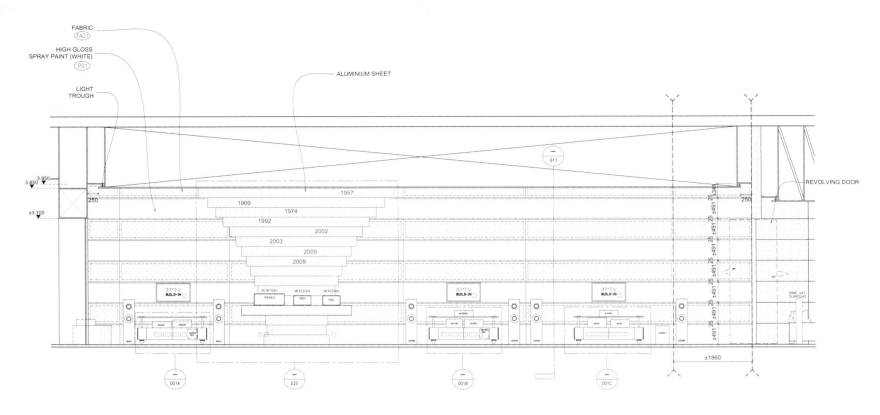

Elevation 3　立面图 3

列布局轻松，共有两种不同形式的陈列台：其中一个的设计灵感来自麦景图音响器材上的大型卷动按钮，另一个则是修长的可丽耐人造石台配水晶玻璃盖，台身的长线设计象征典型的高保真视听器材上的指针。

店内设有两个影院，分别是主影院与迷你影院。主影院里装设麦景图最昂贵的招牌 2K 音响系统，迷你影院则配备时尚的家庭影院系统。与主影院相比，迷你影院的格局较灵活与开敞。

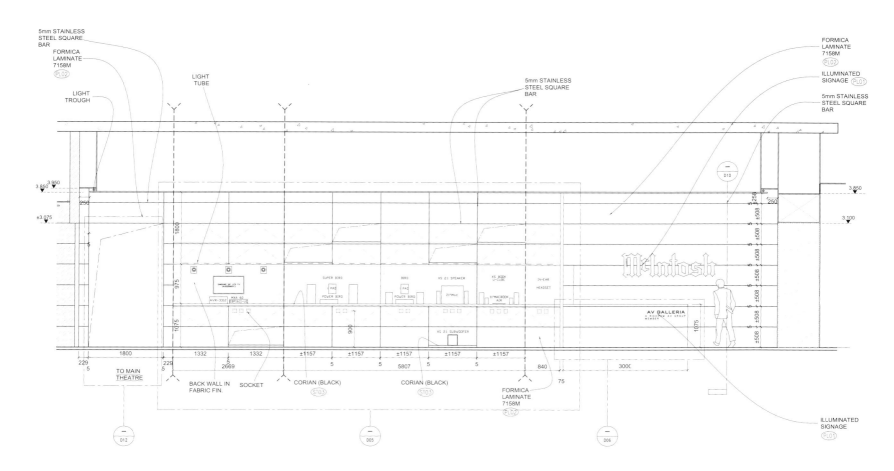

Elevation 4　立面图 4

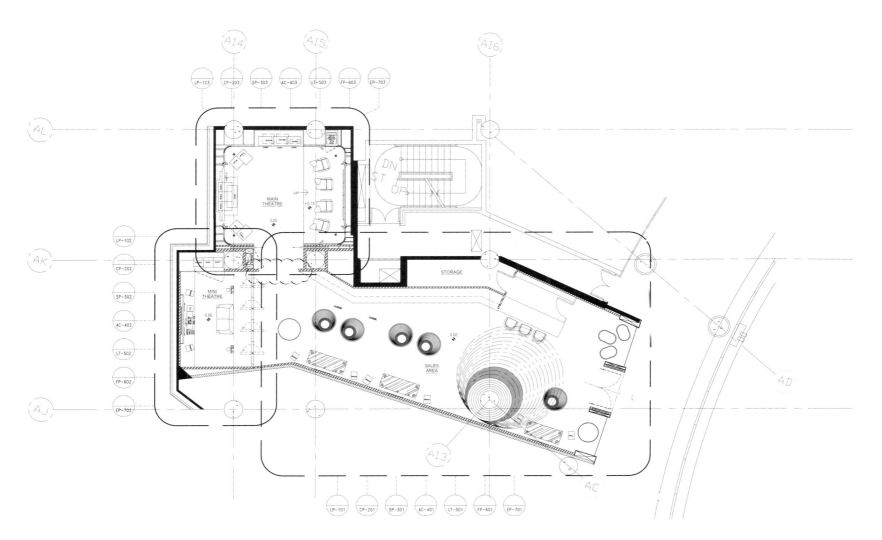

Plan　平面图

O₂ Live Concept Store
O₂ 生活概念店

Design agency : Hartmannvonsiebenthal GmbH
Location : Berlin, Germany
Area : 350 m²
Photography : André Müller

设计单位：Hartmannvonsiebenthal 设计公司
项目地点：德国柏林
项目面积：350 平方米
摄影：安德烈·米勒

The O₂ Live Concept Store presents the O₂ brand's innovative power in one of Berlin's premium retail locations. The aim is to put the customer at the center, translating the company's cross channel-strategy into space. A vivid shop concept puts the focus clearly on experience, relevant services and mobile consultancy.

The O₂ Live concept adjusts to the customers' needs and expectations with strong focus on service and comfort. Local color meets urban culture, Technological innovation, natural materials and handcrafted furnishing make the interior design unique. Illustrations as well as wood create a warm atmosphere. The key objective is to make customers return regularly, anticipating new discoveries as part of the experience. As a result, versatile event schedules with concerts, workshops, cinema, etc., are built up here.

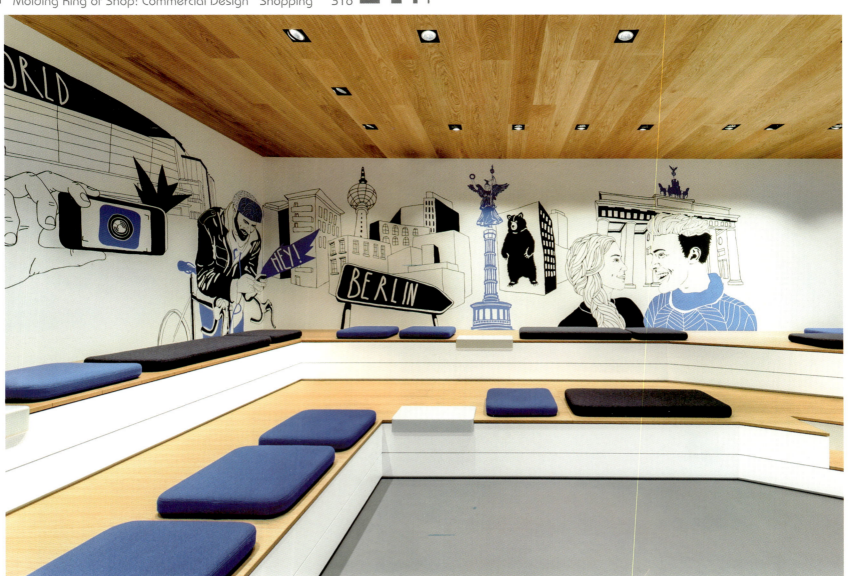

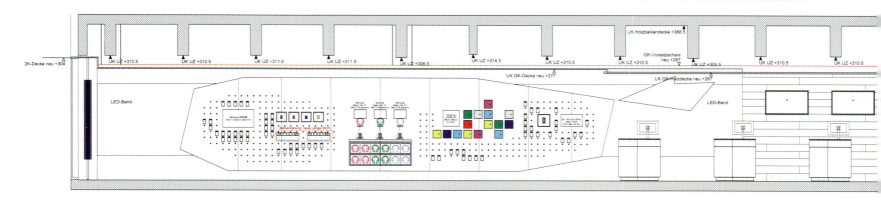

Elevation 1 立面图 1

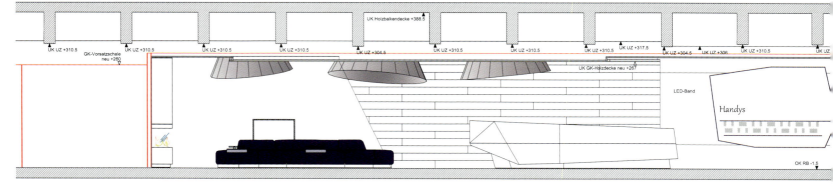

Elevation 2 立面图 2

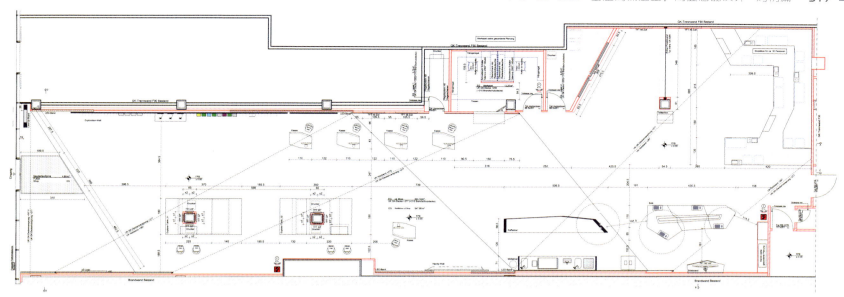

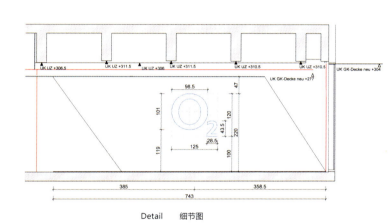

Plan 平面图

Detail 细节图

O₂生活概念店带着其独有的品牌创新力,出现在柏林的一个高档零售区内。设计目的是以顾客为中心,将跨渠道策略融入空间中。该店的设计理念主要专注于移动设备的体验、咨询以及相关服务。

该概念店的设计要满足消费者对服务和舒适度的要求。通过科技创新采用天然材料、手工家具,融合当地色彩和文化,打造出独特的室内空间设计。墙壁上的图画和木材打造出温暖的感觉。主要目的是让客户能够定期回访,每次都能有全新的体验和认知。为此,除了提供全面的服务,O₂生活概念店还将提供音乐会、工作坊、影院等功能空间。

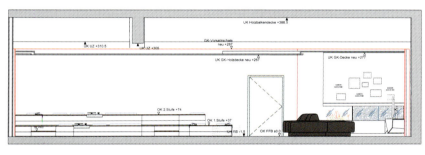

Elevation 3 立面图 3

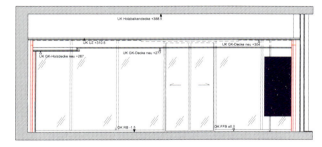

Elevation 4 立面图 4

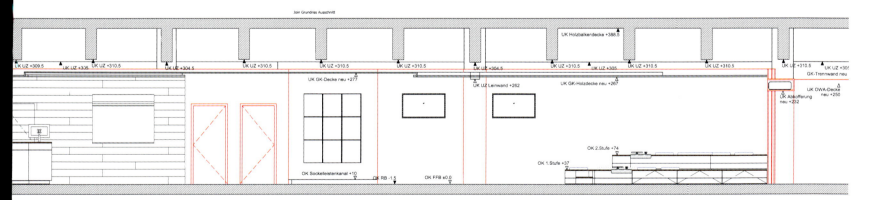

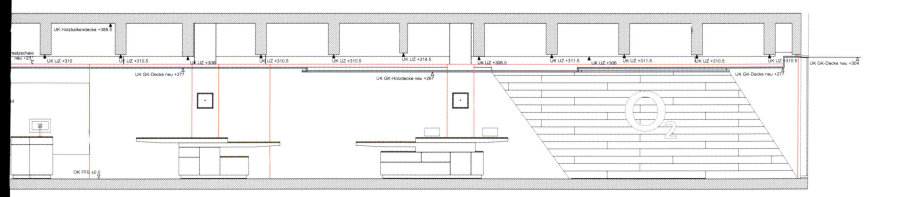

图书在版编目(CIP)数据

塑造商铺之王：商业店面设计. 购物篇 / 深圳市海阅通文化传播有限公司主编. -- 南京：江苏凤凰科学技术出版社，2015.3
 ISBN 978-7-5537-1532-2

Ⅰ.①塑… Ⅱ.①深… Ⅲ.①购物中心—室外装饰—建筑设计②购物中心—室内装饰设计 Ⅳ.①TU247

中国版本图书馆CIP数据核字(2014)第250916号

塑造商铺之王：商业店面设计　购物篇

主　　　编：	深圳市海阅通文化传播有限公司
项 目 策 划：	凤凰空间/彭娜
责 任 编 辑：	刘屹立
特 约 编 辑：	赵　萌
出 版 发 行：	凤凰出版传媒股份有限公司
	江苏凤凰科学技术出版社
出版社地址：	南京市湖南路1号A楼，邮编：210009
出版社网址：	http://www.pspress.cn
总 　经 　销：	天津凤凰空间文化传媒有限公司
总经销网址：	http://www.ifengspace.cn
经　　　销：	全国新华书店
印　　　刷：	深圳市新视线印务有限公司
开　　　本：	889 mm×1094 mm　1/16
印　　　张：	20
字　　　数：	208 000
版　　　次：	2015年3月第1版
印　　　次：	2015年3月第1次印刷
标 准 书 号：	ISBN 978-7-5537-1532-2
定　　　价：	358.00元（精）

图书如有印装质量问题，可随时向销售部调换（电话：022-87893668）。